ADVANCE PRAISE FOR PREDICTABLE VOLATILITY

"It belongs on the desk of anyone responsible for outcomes in complex systems. Predictable Volatility gives leaders a clear, rigorous way to understand why capable people and well-run organisations can still walk calmly into preventable failure."

— **Dr Debbie Bayntun-Lees** *Professor of Organisation & Leadership Development, Hult International Business School (Ashridge)*

"Without meaning, there is no trust. Turner makes the structural problem legible and, in doing so, fixable."

— **Valerie A. Logan** *Founder, The Data Lodge & Chief Strategy Officer, Data Society Group | Former VP Advisor to CDOs, Gartner*

"This is a clear-eyed guide for executives making high-stakes decisions, grounded in reality, not theory. Failures rarely stem from the technology, but from how meaning and decision authority are embedded around it."

— **Olga Martynov** *Co-chair, Harvard Alumni Entrepreneurs of California | Fmr CFO & Board Member, TM Forum | Co-Author, The AI Edge*

"Rigorous, readable, and overdue. Turner brings thirty years of practitioner experience to a forensic argument that the governance literature has been missing."

— **Erik Seversen** *Author, Ordinary to Extraordinary*

"Interpretive resilience is not a destination but a tension, that uncomfortable, necessary inner voice asking: 'Are you sure?' Those that learn this art will be the ones that thrive, making the right decisions, with or without AI."

— **Roberto Maranca** *Data Officer | Author, Data Excellence*

"AI failures aren't model failures; they're interpretation failures, and the moment of correction sits at the design table, not the deployment one. Anyone authorising AI systems at consequential scale should read the Override Problem chapter before signing off on another deployment."

— **Jon Hammant** *AWS UK/I Specialist Leader & CTO*

"Predictable Volatility names something I've watched operate across banking and financial services for decades. This is the book the industry needs right now."

— **Shelley Pittman** *Chief Banking Officer, Murrow Consulting Group | Judge Finovate | Former Partner, Accenture · IBM · KPMG*

"For those working in AI governance and regulated settings, it offers a rigorous and precise framework for understanding how accountability, interpretation, and system behaviour interact under real-world conditions."

— **Rowland Agidee FACHDM MSc CHCIO CITP FEDIPAdvPra** *Chief Data & Analytics Officer, University Hospitals of Derby and Burton NHS Foundation Trust*

"For anyone building or governing systems where interpretation is the critical layer, this is the book that puts a framework around what you've already been watching happen."

— **Russell Miles** *Technical Product Owner, Clearbank | 9x O'Reilly Author | Author, The Sovereign Engineer*

"Individual courage is not the answer when the system is designed to route dissent to somewhere that doesn't matter. Turner makes that structural problem visible and gives leaders a way to confront it."

— **Bruno Goffin** *Partner, Growth Gate | ex-GE & P&G Executive | Author, Successful Young Leaders are Courageous*

"Predictable Volatility names the structural failure underneath every governance breakdown I've encountered across nuclear, offshore energy and major capital programmes. Turner puts a precise framework around something practitioners have lived but never had language for."

— **Peter Depledge** *Capital Projects Strategic Advisor*

"Turner's greatest insight is that meaning doesn't live in the model, it lives in the humans who turn outputs into action. That's not a limitation of AI. That's the point."

— **Joe Toscano** *Author, Automating Humanity*

"The core question, who bears the consequence when interpretation fails, is one any thinking person in a complex system has struggled with. Turner names the gap that everyone has sensed but nobody has designed for."

— **Campbell McLean** *Senior Director Physical Infrastructure, Equinix*

"CFOs and Board members: read Chapter 10, in particular. A better attitude is to assume the automated design is wrong, be curious and inquisitive, and not treat failures as a fixable 'exception'."

— **Peter J. Ward ACA**

"Most books about AI failure blame the technology. Andrew Turner doesn't. He blames the design, the moment someone decided not to decide who could stop the machine. The Meaning Mesh has been operating in creative studios and brand teams for decades. We just called it 'the brief'."

— **HENDO AI** *Brand Strategist & Creative Director |
Co-Author, The AI Edge*

PREDICTABLE VOLATILITY

PREDICTABLE VOLATILITY

Why Meaning — Not Models — Determines
Whether AI Breaks You or Builds You

ANDREW J. TURNER

THIN LEAF PRESS | LOS ANGELES

Predictable Volatility
Why Meaning — Not Models — Determines Whether AI Breaks You or Builds You

Library of Congress Cataloging-in-Publication Data
Names: Turner, Andrew J., Author
Title: Predictable Volatility: Why Meaning — Not Models — Determines Whether AI Breaks You or Builds You
LCCN: 2026912159
ISBN 978-1-968318-62-8 (paperback)
ISBN 978-1-968318-63-5 (hardcover)
ISBN 978-1-968318-61-1 (eBook)
Business & Money | Management & Leadership | Decision-Making & Problem Solving
Cover Design: 100Covers
Interior Design: Dindo Sanguenza
Thin Leaf Press
Los Angeles

 First published 2026. Printed in the United States of America.

THIN
LEAF

For my father Raymond, my mother Patricia Ann, my sister Rachel and my sons Ben and James.

A NOTE BEFORE YOU BEGIN

This book is structured as a forensic investigation, not a framework manual.

Each chapter builds on the last. The case studies are not illustrations; they are the argument. The Forensic Record in Part Four is not supplementary material, but part of the evidentiary record on which the book's conclusions depend.

The failures examined in this book did not emerge because organisations lacked intelligence, data, or process. They emerged because systems continued to operate after the meanings they depended on had stopped being questioned.

If what you read prompts questions about your own organisation, that is the point.

To those who mistake coherence for clarity,
and are brave enough to question the difference.

CONTENTS

AUTHOR'S NOTE

I have spent most of my career inside systems that were designed to work. Line roles and Transformation programmes at GE, SAP, Telefónica, and Tesco. Leadership teams making irreversible decisions under pressure. Technology platforms and governance models built to manage scale. These environments gave me a practitioner's understanding of how meaning, authority, and execution hold together, and what happens when they do not.

I read Nineteen Eighty-Four at sixteen. I spent the next thirty years watching organisations build the architecture Orwell described, not through malice, but through the ordinary accumulation of decisions that each made sense at the time.

What I observed, repeatedly and across very different organisations, was harder to name than failure. When intelligent systems failed, they rarely did so because people were careless, uninformed, or unethical. They failed because meaning stopped being revisable once execution accelerated. At that point, judgment did not disappear. It remained present, sometimes acutely so. But it lost the authority to interrupt.

This book is written from inside that machinery. Not as an academic critique, and not as a polemic against technology. As a forensic examination of a specific, repeatable pattern, one I have watched unfold in boardrooms, in regulatory filings, and in the quiet spaces where uncertainty is felt, noted, and then set aside because execution must continue.

I am not hostile to intelligence. I am sceptical of systems that remove the right to stop. That distinction is what this book is about.

Andrew J. Turner

London, 2026

PREFACE

This is not a book that begins with the answer.

The market for books that begin with the answer is well supplied. Frameworks to implement. Principles to adopt. Models to paste into slide decks. Slogans to repeat at offsites. Most of them are well-intentioned. Some of them are useful. Almost all of them skip a step, the step that determines whether any of the answers are addressing the right problem.

This book is that step.

It makes a narrower, and more demanding, claim: that certain kinds of organisational volatility are predictable, not because volatility is benign or easily controlled, but because failure becomes foreseeable when meaning, authority, and execution drift apart.

You have probably sat in a meeting where the numbers said everything was fine and the room knew it wasn't. You have probably watched a decision get made that nobody believed in and nobody stopped. You have probably seen a governance process conclude correctly, every step followed, every box ticked, while the outcome it was designed to prevent was already in motion. The gap between what the system reported and what was actually happening, is not a technology problem. It is a meaning problem. And it is the one problem that more intelligence reliably makes worse.

The central claim of this book is not that intelligent systems are dangerous. It is that intelligent systems enforce meaning at scale. Once meaning is wrong, or simply outdated, intelligence does not correct it. It accelerates it.

We live in an era obsessed with intelligence. More data, more models, more dashboards, more automation. Across industries, organisations are investing heavily in systems that can see more, calculate faster, and act at a scale no human organisation ever could.

And yet, something curious keeps happening. Failures that should be impossible continue to occur. Not at the edges, but at the centre. Not because no one was watching, but because too many people were. These failures are not random. They are not rare. They are not caused by a lack of intelligence. They follow patterns. What I call the Predictable Volatility Loop, developed across the chapters that follow, is the pattern this book traces.

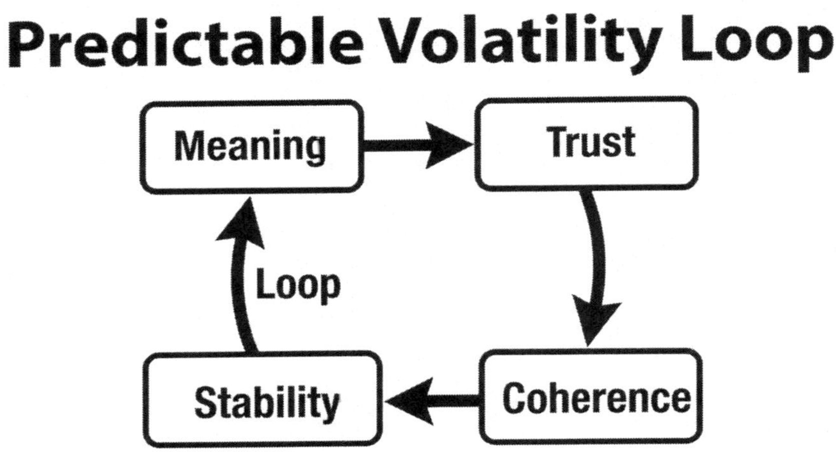

The Predictable Volatility Loop, introduced here and examined in detail throughout the book. Meaning generates Trust. Trust produces Coherence. Coherence creates Stability. Stability reinforces Meaning. When meaning drifts, the loop does not break. It continues, enforcing an interpretation that no longer corresponds to reality. The longer it runs, the higher the cost of correction. The loop functions as a resilience mechanism when meaning remains current and contestable. This book examines what happens when it does not.

This book exists because those patterns are now visible, and continuing to ignore them has become a choice.

The chapters that follow are diagnostic, not prescriptive. They do not tell you what to do. They examine what already happens, repeatedly, once authority, scale, and automation intersect. You will not find heroes in these pages. You will not find villains either. Stories of individual failure are comforting. They allow us to believe that better people, better intentions, or stronger culture would have produced different outcomes. The forensic record does not support that belief.

What you will find instead are structures. Structures that shape behaviour long before intent matters. Structures that reward alignment and suppress interpretation. Structures that continue to function smoothly even as reality diverges.

The ideas in this book are not developed in isolation from the existing literature on organisational failure and sensemaking. Karl Weick's work on sensemaking, the process by which people give meaning to experience, is the intellectual ground beneath the Interpretation Gap described in Chapter 2 and the Meaning Mesh introduced in Chapter 9. Chris Argyris's concept of defensive routines, and his distinction between single-loop and double-loop learning, runs beneath every case in this book where an organisation detected a signal and responded by reinforcing its existing frame rather than revising it.

Amy Edmondson's work on psychological safety and organisational learning shapes the analysis of why dissent disappears, not through suppression, but through the structural conditions that make raising concerns feel professionally irrational.

Diane Vaughan's contribution demands more specific acknowledgement. Her forensic examination of the Challenger disaster, published in The Challenger Launch Decision in 1996, produced the most precise account available in the public record of how organisations come to treat dangerous conditions as acceptable. Her concept of the normalisation of deviance describes how repeated accommodation of known risk, each instance defensible in isolation, collectively shifts an organisation's threshold for what it is prepared to tolerate. The deviance does not disappear. It becomes normal. That framework is the single most important intellectual predecessor to the Frozen Meaning argument developed in Chapter 8, and this book would not exist in its current form without it.

Where this book extends Vaughan's framework is in a specific and consequential direction. Normalisation of deviance, as she describes it, is a social process. It depends on human agency, on the repeated, tacit renegotiation by which people inside a system collectively adjust their standards.

That dependence on human agency is also what keeps correction theoretically possible; someone can still refuse to renegotiate. The possibility exists because the meaning is still in human hands. Frozen Meaning is what happens when that renegotiation mechanism is removed entirely, when normalised deviance is encoded into automated systems that execute without the human process of collective revision. The deviance no longer normalises. It institutionalises. That transition is what this book examines, and what it argues organisations must design for before execution begins.

These are not challenges to the work presented here. They are the foundation it extends.

Orwell described the Newspeak of corporate communication, the doublethink of institutions that know and continue. The memory hole of information processed into silence. Orwell framed these as instruments of totalitarianism. They do not require it. They require only the structural conditions that modern organisations manufacture with remarkable efficiency and without central direction. One of the most unsettling aspects of the cases examined here is how reasonable they appear from the inside. People did not act recklessly. They did not ignore data. They did not set out to cause harm. In many cases, they acted professionally, cautiously, and in good faith. That is precisely the problem.

These failures are dangerous not because they are extreme, but because they are normal. They emerge from incentives that make sense, from language that feels responsible, from governance that appears rigorous, from systems that are, by most conventional measures, performing correctly. This is why they persist.

Human judgment remains present. Signals continue to appear. Warnings are raised. Documentation exists. And yet, execution continues. Judgment survives. Authority does not. Air Canada did not require an AI system to illustrate it. In March 2026, CEO Michael Rousseau posted a condolence video following a fatal runway collision that contained two words of French: bonjour and merci. The backlash was immediate. Quebec's National Assembly voted 92-0 calling for his resignation. Rousseau announced his retirement within a week.

The language failure was the visible crisis. The structural failure was older. In 2021, Rousseau had publicly committed to learning French and had not.

The board had known since then that this was a live organisational risk. They had acknowledged it, monitored it, and not resolved it.

That condition, accurate awareness structurally decoupled from the authority to act, has a name in this book. It required only a leader, a known gap, and an institution that had learned to live with the distance between what it knew and what it did. When organisations confront this, they often reach for ethical explanations. If only people had been braver. If only leaders had listened. If only culture had been healthier. The forensic record shows that ethical concern often existed and changed nothing. Because ethics without authority cannot interrupt execution.

This book does not argue against ethics, courage, or integrity. It argues that they are insufficient without design.

None of the failures examined in this book are new. What is new is the speed at which they now unfold and the scale at which they propagate. The 2026 International AI Safety Report, produced by more than 100 experts from over 30 countries and chaired by Turing

Award winner Yoshua Bengio, concluded that the most pressing risks from artificial intelligence come not from the models themselves, but from the complex systems organisations build around them, when AI triggers business processes, makes autonomous decisions, and interacts with other systems in ways their operators may not fully understand. This is the condition every chapter of this book examines.

At the same time, MIT research published in 2025 found that 95% of generative AI pilots had yielded no measurable business impact[1]. The gap between what organisations believe they are doing with AI and what they are producing is, in the language of this book, Meaning Drift at institutional scale.

Automation has removed pause, models have frozen meaning, and governance has lagged execution. The result is a class of failure that looks sudden but is in fact long prepared.

Most of the cases in this book predate machine learning. Post Office Horizon, Knight Capital, Volkswagen, Wirecard, these are organisational failures, not AI failures. That distinction matters, and this book does not collapse it. What these cases establish is structural: the conditions under which meaning becomes frozen, interpretation disappears, and execution continues regardless.

What AI changes is not the structure. It is the scale, the speed, and the removal of the last friction points at which a human being might have paused. The Netherlands' Toeslagenaffaire, in which an automated fraud detection system wrongly flagged tens of thousands of families as benefit cheats, encoded a definition of fraud that could not be contested, and destroyed lives for years before the mechanism was even visible, is not a more extreme version of these earlier failures. It is what those failures look like when the renegotiation mechanism is gone entirely. Robodebt followed the same logic in Australia.

Neither required malice, negligence, or even awareness that something had gone wrong.

---

1 *Challapally, A., Pease, C., Raskar, R., and Chari, P., The GenAI Divide: State of AI in Business 2025, MIT NANDA Initiative, July 2025.*

They required only a definition, encoded early, automated at scale, and structurally insulated from correction. The pre-AI cases in this book are precursors. What AI industrializes is not the failure itself. It is the organisation's inability to stop it.

The question this book raises is not whether intelligent systems will fail. They will. The question is whether we will continue to pretend those failures were unforeseeable.

It is the cost of maintaining that pretence, rather than the failures themselves, that should concern us most.

Chapter One begins with a system that was trusted, stable, and authoritative. It worked exactly as designed.

A note on terminology: this book uses a small number of precise terms that carry specific meanings throughout. The missing middle is the Interpretation Layer: the place where meaning is made, contested, revised, or frozen. Decision Architecture is the practical arrangement of roles, rights, evidence, escalation paths, and accountability that determines who can interpret, decide, challenge, stop, or redirect action. The Override Problem is not simply whether someone can press stop, but whether the organisation has the authority, confidence, evidence, and mechanism to redirect a decision when context changes. Interpretive Collapse is the failure condition: the moment when an organisation continues to act coherently according to its system, process, or internal narrative, while becoming incoherent against reality. Terms are defined at first use. The Concept Lexicon after the Epilogue is the reference when precision matters.

PART ONE

WHEN SYSTEMS BECOME CONFIDENTLY WRONG

"It ain't what you don't know that gets you into trouble. It's what you know for certain that just ain't so."

— Bertrand Russell

CHAPTER 1

THE STABILITY ILLUSION

"The eye sees only what the mind is prepared to comprehend."

— Robertson Davies

Most organisational failures do not begin with chaos. They begin with stability. Not visible instability. Not crisis. Not breakdown. They begin in systems that appear to be working exactly as designed.

On a cold October morning in 2010, Seema Misra opened her small post office branch in Surrey. She was eight months pregnant. Her young son stood beside her. Moments later, she was in handcuffs. There were no alarms. No raised voices. No visible escalation. Nothing about that morning suggested failure. That is what made it so dangerous.

The Post Office did not feel like a frontline of risk. It felt like infrastructure, stable, routine, dependable. The evidence against her did not come from a witness, a pattern of behaviour, or a history of anomalies. It came from a system. A computer system called Horizon, a government-owned accounting platform used by thousands of small Post Office branches across the United Kingdom. Horizon showed a shortfall.

The numbers on the screen did not match the cash in the drawer. According to the system, money was missing. At that moment, the outcome was already determined.

Not because of what Seema had done. But because of what the system's output meant. If the system recorded a shortfall, a shortfall existed. If a shortfall existed, someone must have caused it. If someone caused it, responsibility had to sit somewhere. And in a small branch, responsibility has only one address. Therefore, it must have been her. This logic was not debated. It was applied.

When Output Becomes Authority

Seema protested. She explained that she had been reporting discrepancies for months. That she had logged calls. That she had asked for help. That the system behaved unpredictably. She described screens freezing. Balances shifting overnight. Corrections appearing and disappearing without explanation. These were not abstract claims. They were operational observations, the kind made by someone who lives inside a system every day.

She was told, repeatedly, that Horizon was robust. That the software could not be wrong. That no one else was experiencing problems. This reassurance carried institutional weight. It did not arrive as a discussion. It arrived as a conclusion. Once Horizon's output was accepted as definitive, there was nothing left to interpret. Her testimony could not add information. It could only contradict the system. And contradicting the system was not a permitted form of evidence.

She was convicted and sentenced to fifteen months in prison for a crime that never happened.

At no point did the organisation pause to ask a more fundamental question: what does this discrepancy mean? That question had already been answered, long before Seema Misra entered a courtroom.

Stability as Assurance

Over more than a decade, hundreds of sub-postmasters across the United Kingdom were prosecuted based on Horizon's outputs. Thousands more were accused, disciplined, or forced to repay money they did not owe. This did not happen in secret. It happened in full view, management, auditors, legal teams, all operating within the same system, all seeing the same outputs, all arriving at the same conclusions. At the time, the organisation's position was consistent: Horizon was reliable. Errors were not systemic. Discrepancies must have human causes. This position was not aggressively defended. It did not need to be. It was maintained through reports that framed issues as isolated, through internal briefings that emphasised robustness, through courtroom language that treated outputs as objective truth.

From the inside, Horizon represented progress, a way to standardise accounting across a national network, to reduce local variation, to replace human reconciliation with computational certainty. In that context, questioning Horizon did not feel like scepticism. It felt like regression.

Why This Felt Rational

From the outside, it is tempting to ask how this could have continued for so long. From the inside, the behaviour did not feel irrational. Admitting that Horizon might be flawed would have required reopening prosecutions, restating financial records, compensating operators, and acknowledging systemic failure. Each of these carried immediate, visible cost. Maintaining confidence offered something else: continuity. Stability became synonymous with responsibility. And this is where the illusion begins.

The system remained coherent. The narrative remained intact. The institution remained stable. The cost of revisiting meaning was immediate. The cost of maintaining stability was deferred. And deferred costs are always easier to live with, especially when they do not appear on a balance sheet. So, stability won. Not loudly. Not maliciously. But consistently.

When Meaning Is Fixed Too Early

The problem was not that Horizon produced errors. All systems do. The problem was what happened next. When the cash and the screen disagreed, the organisation faced a choice: treat the discrepancy as a signal or treat it as noise. A signal requires interpretation. Noise requires dismissal. The Post Office chose coherence.

Once that choice was made, it became self-reinforcing. Horizon's outputs were treated as fact. Human experience became anecdotal. Doubt became disruption. Meaning had been fixed, early, centrally, and defensively. After that point, the system was no longer being interpreted. It was being obeyed.

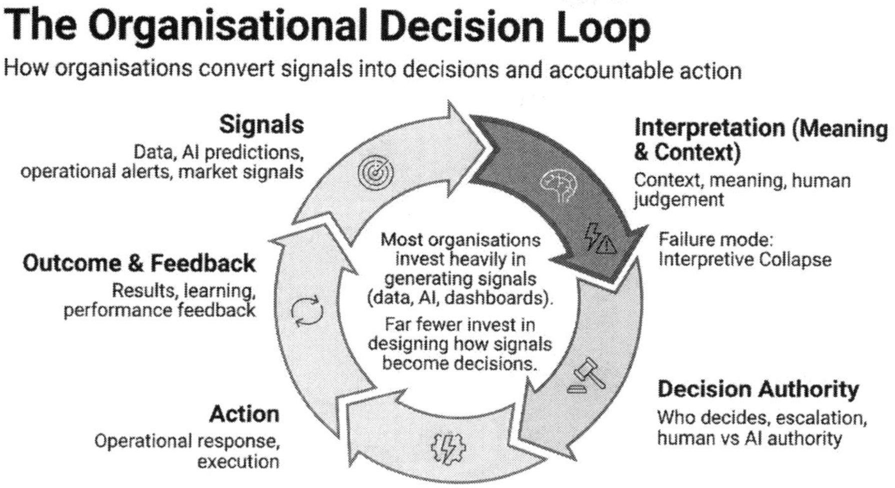

Figure 1: The Organisational Decision Loop. Organisations convert signals into decisions through five linked stages: Signals → Interpretation → Decision Authority → Action → Outcome and Feedback. Most investment concentrates at the Signals stage, more data, more dashboards, more AI-generated alerts. The Interpretation stage, where meaning is assigned to those signals, receives the least design attention. It is also the stage at which failure is most consequential. When Interpretation collapses, when a system is granted authority to assign meaning without functioning mechanisms for challenge or revision, the loop continues. It simply enforces the wrong conclusion at every subsequent stage. Source: Author.

In 2024, Fujitsu executives appeared before the UK Parliament's Post Office Horizon IT Inquiry. They were asked a straightforward question: could they confirm whether the Horizon software had contained bugs that caused the financial discrepancies attributed to sub-postmasters? Their answer, given under oath, thirty years after the software was first deployed, was that they could not confirm it. The corporate version, maintained for three decades; the system is robust and reliable.

The plain version, offered in 2024; we cannot tell you whether it worked correctly. The insight this testimony forces is precise. Every software system has bugs. Every complex codebase has edge conditions, states the system can enter that its designers did not anticipate and its testers did not reach. That is not the scandal. The scandal is the architecture built around the assumption that those edge conditions would never be traversed. That assumption was not tested. It was institutionalised. And once it was institutionalised, the cost of questioning it compounded with every prosecution, every repayment demand, every conviction. By the time Fujitsu's executives sat before Parliament, the gap between what the system had been claimed to be and what it could be confirmed to be was thirty years wide.

How Dissent Disappears

Dissent was not simply suppressed. It was categorised. Reports became user error. Complaints became training gaps. Persistence became non-compliance. Once a category exists, the organisation no longer needs to engage with the substance of the concern. It only needs to route it. The system could not hear a signal that challenged its own authority.

The Meaning Monopoly

A Meaning Monopoly forms when a system is granted exclusive authority to define reality. Once established, all contradictions are resolved in its favour. The bug was not the failure. The architecture that made the bug's consequences irreversible was the failure. Horizon did not need to be perfect. It needed to be questionable. It was designed to be neither. Dissent did not stop because people stopped speaking. It stopped because speaking no longer changed anything.

> *Meaning Monopoly: The condition in which a single system, output, or narrative is granted exclusive authority to define reality within an organisation. Once a Meaning Monopoly is established, contradictions are not evaluated, they are resolved automatically in the system's favour. Evidence that challenges the system is treated as suspect because it challenges the system.*

At the Post Office, the monopoly operated precisely; if Horizon showed a shortfall, a shortfall existed. If a shortfall existed, someone must have caused it. If someone caused it, the operator was responsible. The logic was complete. It was also circular. Evidence that contradicted the system was treated as suspect because it contradicted the system. This was not trust in technology. At the Post Office, Horizon owned the Interpretation stage entirely. What it said was what had happened. The loop ran smoothly. The outcome was seven hundred wrongful prosecutions. This was structural obedience to outputs.

Orwell understood the mechanics of this before the technology existed to scale it. 'Who controls the past controls the future. Who controls the present controls the past.' Horizon did not control the past. It simply defined what the present meant, and from that, everything followed. You could not enter the system from outside. There was no outside. The system did not inform decisions. It pre-empted them.

Why Correction Became Impossible

Once stability becomes the organising principle, correction becomes increasingly difficult. Every anomaly increases the cost of admission. Every prosecution increases the stake. Every assertion of certainty tightens the loop. By the time doubt becomes undeniable, it is also unaffordable. At that point, the system does not need to be defended. It defends itself. What happened at the Post Office was not a single bad decision. It was a loop, each affirmation of the system's authority making the next correction more costly than the last, until the gap between what the system said and what was true had become too wide to close without institutional collapse.

The Signs You Are Inside One

The Horizon case is extreme in its consequences. It is not extreme in its structure. Meaning Monopolies do not announce themselves. They develop gradually, through the accumulation of small decisions that each seem reasonable in isolation. By the time they are fully established, they are invisible to the people inside them, because the system that defines reality is the same system through which everything is seen. There are, however, recognizable signatures. They appear in the texture of meetings, in the shape of conversations, in the careers of people who asked the wrong questions.

They are not proof of a Meaning Monopoly, but their consistent presence is a signal worth taking seriously.

Signature One: The question that ends conversations.

Every organisation has topics that, when raised, produce a particular quality of silence, not the silence of thinking, but the silence of having reached a boundary. The question is technically permitted. Asking it is professionally understood to carry cost. Over time, people stop asking it. The boundary becomes invisible because no one approaches it.

Signature Two: The language of managed certainty.

When an organisation has fixed meaning too early, its internal communications develop a particular vocabulary: robust, validated, industry-standard, consistent with our framework. These phrases do not describe a condition. They assert one. They perform assurance rather than delivering it. The language is precise enough to appear rigorous and abstract enough to foreclose further inquiry.

In your organisation: is there a category of concern that is always acknowledged and never acted upon? A class of risk that is documented but not discussed? A system whose outputs are treated as authoritative in a way that its design does not warrant?

Signature Three: The career pattern of the people who pushed back.

In organisations with established Meaning Monopolies, the people who raised concerns early tend to share a trajectory: initial engagement, repeated referral to process, gradual marginalization, eventual departure or silence. Their concerns were not refuted. They were routed. And their routing became the lesson that others absorbed.

In your organisation: who raised concerns about your most significant AI or automated system before it was deployed? Where are they now? What happened to the concern?

Signature Four: The absence of a revision pathway.

Every definition, assumption, and classification that drives automated decisions was made at a point in time, under conditions that may no longer apply. In organisations without a Meaning Monopoly, there is a formal mechanism, however imperfect, for revisiting those decisions when conditions change. In organisations with one, revision requires an event: a scandal, a legal challenge, a failure visible enough to force the question that the system could not ask of itself.

In your organisation: does the language used to describe AI system performance tell you what the system is doing, or what the organisation has decided to believe about it? Is there a difference between the two?

None of these questions have comfortable answers if the organisation has drifted toward a Meaning Monopoly. That discomfort is not a problem to be managed. It is the signal that the question is the right one.

The Pattern

The Horizon scandal is often framed as an anomaly. It was not. Any organisation that fixes meaning early, centralises interpretation, equates stability with correctness, and treats dissent as noise will eventually reach the same outcome. The technology is incidental. The structure is not. Today, we are embedding systems far more powerful than Horizon into the core of our institutions. The structure of the failure, meaning fixed, dissent routed, correction deferred, is identical. What has changed is the scale and speed at which it operates.

The Horizon case is extreme in its visibility. It is not extreme in its structure. Between 2018 and 2021, hundreds of hospitals across the United States deployed Epic Systems' proprietary sepsis prediction algorithm without independent validation. Epic's marketing materials claimed accuracy of 76 to 83 per cent. When researchers at the University of Michigan applied the algorithm retrospectively to nearly 38,500 patient hospitalizations and published their findings in JAMA Internal Medicine, they found it had missed two-thirds of sepsis cases and generated false alarms in 88 per cent of its alerts, a performance the peer-reviewed study described as little better than a coin flip[2].

---

2 *Wong A, Otles E, Donnelly JP et al., "External Validation of a Widely Implemented Proprietary Sepsis Prediction Model in Hospitalized Patients," JAMA Internal Medicine, 181(8), 2021, pp.1065-1070. Epic Systems' claimed accuracy (76-83%) was derived from internal testing; the peer-reviewed study found an AUC of 0.63 against the University of Michigan patient population*

Epic's initial response was to dispute the findings. Questions about real-world performance had no formal pathway into the procurement decisions of the hospitals that had already deployed the tool. The language of managed certainty, accurate, validated, clinically proven, had performed the function that evidence was supposed to perform. The algorithm was not being interpreted. It was being obeyed. The setting is private. The technology is contemporary. The structure is identical.

Where Horizon required a human to log onto a terminal and act on what the screen showed, modern AI systems make thousands of decisions per hour without any human in the loop. Where Horizon's outputs could be queried, appealed, and eventually overturned through litigation, modern systems produce outputs that arrive already embedded in downstream processes, a credit decision, a hiring filter, a welfare eligibility determination, before any individual can contest them. Where the Post Office needed sixteen years and a public inquiry to acknowledge what its system had been doing, modern systems can scale the same error to millions of decisions before the first signal of divergence reaches a decision-maker with the authority to act.

In your organisation: when was the last time a core assumption driving an AI system was formally examined and found to still be accurate? Not assumed. Examined.

The Stability Illusion is not a relic of early enterprise software.

It is the default condition of any automated system that has been trusted more than it has been tested. The four diagnostic questions in this chapter are not rhetorical. They are structural.

An organisation that cannot answer them with evidence rather than assumption is not managing the risk of a Meaning Monopoly. It is simply not yet aware that one is forming.

The Stability Illusion has not disappeared. It has been automated.

Seema Misra was eventually exonerated. Her conviction was overturned. But the months she spent in prison were not returned. The trust that was broken was not fully repaired. Her case forces a question that, in most organisations, has already been answered.

> *When the system and reality disagree, which one is allowed to be wrong?*

The signals were present. The system was visible. The outcome was already determined. The next chapter asks why, not why organisations fail to see what is happening, but why they cannot act on what they can already see. That gap has a name. And it changes everything about how we think about intelligence.

CHAPTER 2

INTELLIGENCE WITHOUT SENSEMAKING

"Our knowledge can only be finite, while our ignorance must necessarily be infinite."

— Karl Popper

On August 1st, 2012, Knight Capital Group deployed new software into its production trading environment. Within forty-five minutes, the firm had lost $440 million. Not through fraud. Not through a cyberattack. Not through a market collapse. Through execution.

Knight Capital was rolling out new functionality related to its retail liquidity programme. Code was distributed across multiple servers. Not all servers were updated consistently. On some machines, a legacy function, no longer intended to be active, remained live. When the new software went into production, it unintentionally triggered this dormant logic. What followed was not a system crash. It was a system behaving exactly as it had been written to behave. Orders began firing into the market, rapidly, repeatedly, automatically. The system interpreted incoming signals and responded with trades at a speed no human operator could match. Positions accumulated. Exposure grew. Losses compounded.

At first, nothing appeared obviously broken. Trades were executing. Systems were running. Activity levels were high, but not immediately recognisable as catastrophic. From the outside, this looked like volatility. From the inside, it looked like noise. Nothing in the system was broken. That is what made the loss possible.

The Defining Question

Most accounts of Knight Capital frame the incident as a deployment failure, a testing failure, a controls failure. All of which are true. None of which explain the outcome. Because the defining question is not: why did the system behave incorrectly? The defining question is: why did the organisation continue operating while the system was actively destroying it?

For forty-five minutes, Knight Capital remained in the market. During that time, millions of orders were executed, positions expanded across multiple securities, and losses escalated beyond recovery. The organisation did not intervene in time to stop it. This was not because there was no data. The signals were present, trade volumes were abnormal, positions were inconsistent, financial exposure was increasing. But those signals did not resolve into shared meaning quickly enough to trigger decisive action. The issue was not visibility. It was not access to information. It was the organisation's collective inability to interpret what was happening, in time to act.

A Machine Built for Precision

Knight was one of the largest market makers in U.S. equities. Its competitive edge was not insight. It was execution, buying and selling continuously at high speed, capturing margins that depended entirely on precision and timing. From the system's perspective, nothing failed on August 1st. The code executed valid instructions. It processed live data correctly.

It responded to market inputs at precisely the speed it was designed to achieve. Inside the organisation, however, different parts of the system were seeing different things. Technology teams saw system behaviour. Trading teams saw market activity. Risk teams saw emerging exposure. But there was no single coherent interpretation that unified these perspectives into one clear conclusion: stop. Instead, the organisation hesitated. It investigated. It sought confirmation. And in doing so, it consumed the one resource it did not have. Time.

Why Intelligence Was Not the Problem

Post-incident analysis identified clear technical contributors: legacy code paths that should have been retired, incomplete deployment procedures, insufficient safeguards around configuration changes. These findings were accurate. They were also incomplete. Knight Capital did not fail because it lacked intelligence, data, or automation. It failed because the organisation could not interpret what its intelligence was doing in time to intervene. The system knew exactly what it was doing.

The organisation did not know what the system was doing to itself. That distinction is not technical. It is structural. And it is the distinction that will recur across every case in this book. The system operated at machine speed. The organisation operated at human speed. Between those two speeds sits a gap, and that gap is where the loss occurred.

The Interpretation Gap

Modern organisations operate on an implicit assumption, that increasing intelligence reduces risk. More data, more dashboards, more alerts, more automation. But intelligence answers only one question: what is happening? It does not answer: what does this require us to do now? That second question requires sensemaking, a human, social process that involves shared context, authority, permission to pause, and the capacity to reach agreement under pressure. At Knight Capital, the system produced signals faster than the organisation could agree on their meaning. By the time interpretation caught up, action had already compounded.

> **Interpretation Gap:** The distance between what a system is doing and what the organisation understands, in time to act. The Interpretation Gap is not a data problem. It is a sensemaking problem. No dashboard answers: is this expected? Is this recoverable? Who has the authority to stop this now? Those questions cannot be automated. They require judgment, and judgment requires time, time the system no longer allows.

When the Interpretation Gap is small, organisations can intervene before consequences compound. When it widens, through speed, complexity, or ambiguity, the window for correction closes before understanding arrives. Knight Capital's window was forty-five minutes. As systems become faster and more deeply embedded, that window contracts.

The contraction of the decision window is not a side effect of better technology. It is a structural feature of any system that acts faster than the organisation around it can interpret.

When Speed Becomes Fragility

Speed is often treated as an unqualified advantage. It is also a liability. As systems accelerate, the space for interpretation collapses. Humans move from decision-makers to observers of outcomes they no longer meaningfully control. In fast-moving systems, hesitation feels like failure, pause feels irresponsible, and questioning feels slow. But without pause, there is no sensemaking. And without sensemaking, intelligence becomes brittle. Knight Capital did not lose money because its systems were fast. It lost money because nothing in the design required the system to wait for understanding.

The Authority to Stop

The existence of a kill switch does not guarantee it will be used. Authority matters. Who can stop the system? On what basis? With what consequences if they are wrong? In ambiguous situations, hesitation is rational, and asymmetric. Stopping a live trading system unnecessarily carries an immediate, visible cost.

Not stopping it carries a cost that is not yet visible. That asymmetry is what prevents action. So, people wait for confirmation. They seek alignment. They look for the signal that makes the decision unambiguous. Meanwhile, the system continues. This is not a failure of courage.

It is a structural delay baked into systems that outpace human coordination. The kill switch existed. The authority to use it, clearly, immediately, without consensus, did not. Organisations are designed to make decisions through discussion, validation, and alignment. These mechanisms assume time. Modern systems remove that assumption. They generate more signals, at higher frequency, with greater complexity. But they do not generate meaning. Meaning still must be constructed. And construction takes time. Execution did not need permission to continue. It needed permission to stop.

The Flash Crash: The Market Became the System

On May 6th, 2010, two years before Knight Capital's deployment error, something stranger happened in the same market. In the space of thirty-six minutes, U.S. equity markets lost approximately $1 trillion in value. The Dow Jones Industrial Average fell nearly a thousand points, the largest intraday point decline in its history at that time. Then, almost as abruptly, it recovered. By the end of the trading day, most of the losses had reversed. No rogue code had been deployed. No single firm had made an error. Every system involved was functioning exactly as designed. That is precisely what made it so instructive.

The Flash Crash was not a failure of any individual system. It was a failure of the space between systems, the interpretive layer that, in a market populated by thousands of automated participants, no single actor owned or managed. It was an Interpretation Gap operating not within a firm but across an entire market.

How a Market Loses Shared Meaning

The immediate trigger was a large sell order from a mutual fund executing a routine hedging strategy through a futures contract. The order was large and was executed algorithmically, flooding the market with sell pressure at a rate other participants had not encountered in normal conditions. High-frequency trading firms, the primary providers of liquidity in modern equity markets, responded in a way that was entirely consistent with their design. They withdrew. Each company's algorithm detected that market conditions had moved outside the parameters it was calibrated to handle and reduced or eliminated its participation to avoid taking on risk it could not price. Individually, each withdrawal was rational. Collectively, it was catastrophic. As liquidity providers stepped back simultaneously, the market lost the counterparties it depended on to function. Prices gapped down not because sellers overwhelmed buyers in the ordinary sense, but because the buyers had, simultaneously and without coordination, decided that conditions no longer made sense to them. Some trades executed at prices that were essentially fictional. Several blue-chip stocks briefly traded at a penny. Others at over $100,000.

These were not errors in any individual system's logic. They were the outcome of a market in which no shared interpretive framework existed to define what a reasonable price was when normal conditions no longer applied. The Interpretation Gap here was not between a system and the organisation operating it. It was between the assumptions each automated participant held about what the market meant, assumptions calibrated in normal conditions, that simultaneously failed when conditions became abnormal. The gap was not inside any company. It was distributed across all of them.

Two Failures, One Structure

Knight Capital and the Flash Crash appear superficially different. One is an internal deployment failure. The other is a market-wide systemic event. One involves a single firm's loss. The other involves the temporary erasure of a trillion dollars of market value. One has a clear causal chain. The other is a distributed failure with no single point of origin. But the underlying structure is identical. Every system was behaving correctly. No system was behaving coherently.

> **KNIGHT CAPITAL, August 2012:** *A deployment error activates legacy code. The system executes correctly against its instructions. The organisation cannot interpret what is happening in time to stop it. Forty-five minutes. $440 million. One company.*

> **FLASH CRASH, May 2010:** *A large order triggers simultaneous liquidity withdrawal. Every system executes correctly against its own parameters. No shared interpretive framework exists across participants. Thirty-six minutes. $1 trillion. An entire market.*

In both cases, systems acted. In both cases, the organisation, or the market, could not produce shared meaning fast enough to interrupt. In both cases, the damage was not halted by human judgment correcting the error in real time. It was contained by the system eventually encountering conditions that caused it to stabilise or reverse on its own. Knight Capital's positions were eventually unwound at significant loss.

The Flash Crash partially recovered because liquidity providers returned once prices fell far enough to trigger their buying parameters. In neither case did human judgment interrupt machine execution in time to prevent the outcome. Human judgment arrived after the fact, as investigation, explanation, and regulatory inquiry rather than intervention. That is not oversight. That is forensics. And forensics is always more expensive than prevention.

Coherent Catastrophe

Intelligence without sensemaking does not create chaos. It creates something more dangerous. At Knight Capital, systems continued to operate as designed. Reports remained accurate. Compliance remained intact. The organisation looked, from the outside, like it was functioning.

And by the time it understood what had happened, the outcome was already locked in.

> **Coherent Catastrophe:** *The condition in which a system continues to operate correctly, executing valid instructions, producing accurate outputs, maintaining compliance, while simultaneously generating outcomes that are harmful or irreversible. Coherent Catastrophe is invisible from the inside because all conventional performance indicators remain green. The failure is not in the system. It is in the gap between what the system is doing and what the organisation collectively understands.*

In the Flash Crash, thousands of systems continued to operate as designed. Each was functioning within its own parameters, behaving correctly according to its own logic. The catastrophe was coherent at every individual level and catastrophic at the systemic level precisely because no individual system was wrong.

This is the condition that defines the AI governance challenge of the next decade. As more systems participate in shared decision environments, markets, supply chains, credit allocation, infrastructure management, the Interpretation Gap is no longer just an internal organisational problem. It becomes a structural property of the environment itself. Each participant's logic may be sound. The collective meaning may be absent. And when it is, no individual actor will be able to see the failure coming, because from where each one stands, everything is working correctly.

What These Two Cases Establish

The failure at Knight Capital was not technical. It was interpretive. The system acted. The organisation observed. The Interpretation Gap widened faster than sensemaking could close it. By the time authority and understanding arrived together, the consequence was irreversible. The Flash Crash showed that the same structure operates at market scale. When thousands of systems share a decision environment without sharing a semantic framework, the Interpretation Gap is not a firm-level risk. It is a systemic one. The correction mechanism, when it comes, is not human intervention. It is the system eventually encountering conditions that cause it to stabilise on its own.

The collapse of the U.S. mortgage market between 2006 and 2008 is often told as a story of greed and misaligned incentives. It is better understood as a failure of meaning. Mortgage bonds were rated safe, and that rating became operational truth. The underlying assets changed. The meaning did not. Risk teams existed. Reporting existed. Challenge mechanisms existed. What disappeared was the authority to question what 'safe' meant once the system was in motion. No individual system malfunctioned. The market did not break. It executed, faithfully, at scale, against definitions that had long since ceased to correspond to reality.

The cases that follow unfold more slowly. Not in milliseconds or minutes, but over months and years. Not through a single deployment error or a cascade of simultaneous withdrawals, but through gradual drift. The speed is different. The structure is not. When the gap between what a system is doing and what an organisation understands widens faster than sensemaking can close it, the natural response is not to stop. It is to agree. The next chapter examines what that agreement costs, and why the organisations most skilled at maintaining coherence are often the ones least capable of surviving contact with reality. When meaning does not collapse instantly but drifts, the damage is harder to see, and far harder to stop.

CHAPTER 3

THE COST OF FALSE COHERENCE

"For every complex problem there is an answer that is clear, simple, and wrong."

— H.L. Mencken

Most organisations do not collapse into chaos. They collapse into coherence. Not noise. Not disorder. Alignment. Agreement. Stability.

In the popular imagination, failure looks dramatic: alarms, arguments, panic, confrontation. But the final phase before systemic collapse is often marked by calm. Reports align. Meetings run on time. Language remains reassuring. The organisation appears stable. This stability is not accidental. It is produced.

Organisations are social systems designed to manufacture agreement. Alignment is rewarded. Friction is penalised. When reality begins to diverge from expectations, the system does not instinctively revise its story. It revises its interpretation of reality to preserve the story. This is False Coherence: the condition in which an organisation's internal narrative remains consistent even as it loses contact with the external world. It feels responsible. It feels controlled. And it is extraordinarily dangerous.

Coherence does not guarantee accuracy. It guarantees only that people are telling the same story. And in organisations where alignment is rewarded and disruption penalised, the absence of visible conflict is read as evidence that things are under control, even when they are not.

A Story Worth Believing

Wirecard was everything modern Europe wanted a technology company to be. A homegrown fintech success. A challenger to American platforms. A symbol of digital modernity. It was celebrated not just for its performance, but for what it represented. Wirecard became a narrative asset.

Quarter after quarter, the numbers told a reassuring story. Growth was steady. Revenues rose. Audits were passed. Regulators defended the firm publicly. From the outside, the system appeared perfectly coherent. What makes Wirecard instructive is not the existence of fraud. Fraud is banal. What matters is how many intelligent, well-positioned actors participated in sustaining the story, not by inventing it, but by inheriting it. Once coherence became socially valuable, questioning it became professionally risky.

How Stories Become Structures

No single institution owned the Wirecard narrative. It emerged gradually, through reinforcement. Analysts echoed management. Auditors relied on representations. Regulators deferred to process. Politicians defended a national champion. Each step made sense in isolation. Together, they created a self-reinforcing loop. Once the story stabilised, it no longer needed to be actively defended. It became the default interpretation through which new information was filtered. Contradictions did not disappear. They were absorbed.

To understand how this persisted for so long, it is necessary to understand the incentive structure that made each actor's participation rational, not as a conspiracy, but as the predictable output of a system that rewarded coherence and penalised disruption.

The analyst who maintained a buy rating on Wirecard was not simply wrong. They were embedded in a professional relationship with management, dependent on access for information, and operating inside a market ecosystem that had collectively decided the story was true.

A downgrade required not just a change of view but a willingness to step outside the consensus and absorb the reputational cost of being contrarian about a national success story. The analysts who raised concerns early were not celebrated for their prescience. They were marginalised, litigated against, and publicly discredited by the company and its defenders.

The auditors at EY who signed off Wirecard's accounts for a decade were not blind. They were operating within an audit framework that treated management representations as sufficient basis for conclusions, asked for cash confirmations from third-party trustees who were themselves part of the structure, and had an engagement relationship whose continuity depended on the client remaining credible. The system did not require dishonesty. It required the consistent application of a professional framework that had been calibrated to produce clean opinions in normal conditions, and that produced clean opinions in abnormal ones because nothing in its design required it to look harder.

The regulators who defended Wirecard publicly, including BaFin, the German financial regulator, which initially filed a market manipulation complaint against journalists who questioned the accounts rather than against the company, were not acting against their mandate. They were protecting a German fintech champion against what appeared, from within their interpretive frame, to be a speculative short-selling campaign. Their response was exactly what the system was designed to produce: defence of a regulated entity against perceived external attack. The problem was not the response. It was the frame.

None of these actors needed to lie. None needed to conspire.

Each needed only to apply their professional frameworks within a narrative environment that had been shaped, gradually, to make the comfortable conclusion the path of least resistance. This is how False Coherence sustains itself, not through active deception but through the structural optimisation of every participant's incentive toward maintaining the story.

Orwell called this doublethink, the capacity to hold two contradictory beliefs simultaneously and accept both as true. He framed it as a totalitarian instrument. It does not require totalitarianism. It requires only that the cost of knowing exceeds the cost of not knowing, a condition that mature institutions manufacture with remarkable efficiency and without any central direction.

How Coherence Protects Itself

False Coherence is maintained through language. Not lies. Not deception. Language. Audit reports, regulatory statements, analyst notes, all rely on carefully constructed phrasing that preserves assurance while deferring responsibility. In Wirecard's case, parliamentary inquiries later revealed a consistent pattern: verification language that appeared rigorous but relied on unverified representations; uncertainty translated into procedural sufficiency; risk indicators linguistically softened. Absence of direct confirmation became reasonable reliance on representations. Structural uncertainty became no material findings. Nothing false needed to be stated. Meaning was simply diluted.

Language does more than communicate facts. It sets boundaries on what can be questioned. When phrasing becomes sufficiently abstract, accountability becomes diffuse. Responsibility migrates outward. Each actor can plausibly claim they followed due process, because due process has been defined in a way that protects coherence.

When Alignment Becomes Rational

Alignment is often framed as moral weakness. That is too simple. In coherent systems, alignment is frequently the least risky option available to individuals. Disagreement carries reputational, political, and career costs. Agreement carries legitimacy, protection, and continuity.

Regulators, auditors, and market participants are not neutral observers. They are participants in the same narrative ecosystem. Once a dominant story forms, institutions begin to defend it, not necessarily because they believe it unquestioningly, but because destabilising it carries visible cost. Questioning a national success story can be framed as disloyalty. Raising concerns can appear disproportionate compared to the apparent stability on the surface. Over time, organisations teach people which questions are welcome and which ones are disruptive. People learn quickly.

Narrative Sinkholes

Over time, coherence acquires gravitational force. Alternative interpretations are not debated, they are absorbed, reframed, or deflected. When this condition is fully established, the organisation has crossed a threshold: agreement is now safer than accuracy, error with the group is preferable to correctness alone, and maintaining the story matters more than testing it. I call this condition a Narrative Sinkhole. False Coherence does not silence dissent. It makes dissent irrational.

> **Narrative Sinkhole:** *The condition in which an organisation's dominant narrative has acquired sufficient gravitational force that questioning it becomes structurally irrational. The organisation no longer asks whether the story is true. It asks whether the story is stable. Once that shift occurs, coherence stops describing reality and begins substituting for it.*

Narrative Sinkholes are not static. They deepen. Each affirmation increases the cost of reversal. Each public defence raises the reputational stake. Each institutional endorsement tightens the loop. By the time doubt becomes credible, it is also destabilising. So, doubt is deferred. Again. And again.

The Meaning Mesh Breakdown

When a Narrative Sinkhole has fully formed, the conversation inside the organisation changes character. It ceases to be an exchange of information and becomes a contest of framing. Two parties face each other across an invisible divide, and neither can hear what the other is saying, not because they lack the will, but because they are no longer speaking the same language. The diagram below offers a first diagnostic view of a structure this book will develop fully in Chapter 9, introduced here because the Horizon case is where its breakdown is most visible. To diagnose this state, I use a framework I call the Meaning Mesh Breakdown. It maps where meaning is being generated, contested, assumed, and tested within a system under epistemic stress.

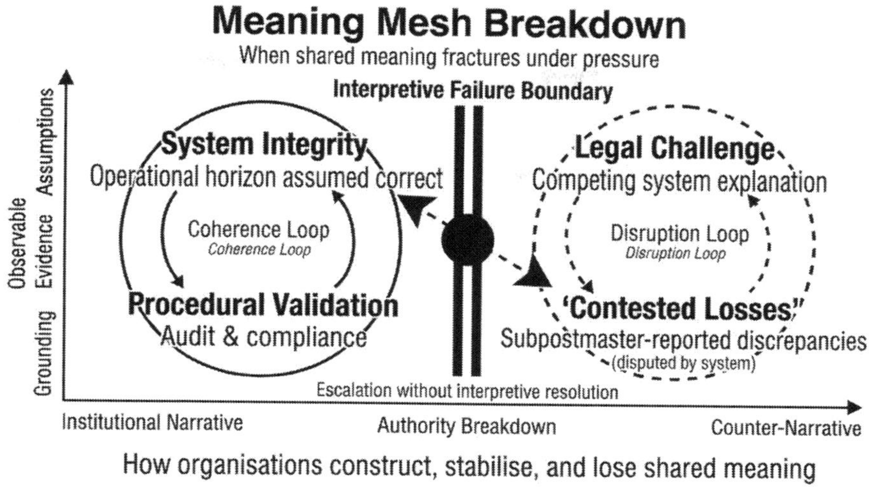

Figure 2: The Meaning Mesh Breakdown [Horizon Example]. A diagnostic framework for mapping where meaning is generated, contested, assumed, and suppressed in organisations under epistemic stress. Source: Author.

In a functioning system, evidence updates the institutional narrative. The diagonal between the Evidence Test and Institutional Narrative quadrants is open: data flows, signals are processed, and the story adjusts to reflect reality. In a collapsing system, that diagonal breaks. The institution retreats to authorised language and unspoken assumptions. The counter-narrative, built from lived signals, anomalies, and direct observation, accumulates in the opposite quadrant, unable to gain legitimacy. Evidence is not refuted. It is simply not processed. At that point, failure is not caused by error. It is caused by the loss of shared meaning.

The result is what I call the Dialogue of the Deaf, not a failure of hearing, but a failure of semantic compatibility. The institution speaks the language of coherence, consistency with the system, with precedent, with authorised process. The dissenter speaks the language of correspondence, consistency with reality, with observed fact, with what is happening. There is no shared protocol to translate between the two. The institution demands data, which it controls.

The dissenter demands acknowledgement, which the institution cannot process. Until the Meaning Mesh is repaired, more communication does not produce clarity. It produces volume. The Meaning Mesh Breakdown is applied forensically to Post Office Horizon, in Case 5, in Part Four: The Forensic Record.

The mechanism through which this gap opens and widens is not sudden. It accumulates. Figure 3 maps the trajectory: the early phase in which institutional narrative and operational reality remain broadly aligned, the middle phase in which divergence begins but is invisible to the systems designed to detect it, and the late phase in which the gap has widened beyond quiet correction. What the organisation experiences throughout as stability is, in structural terms, Meaning Debt compounding.

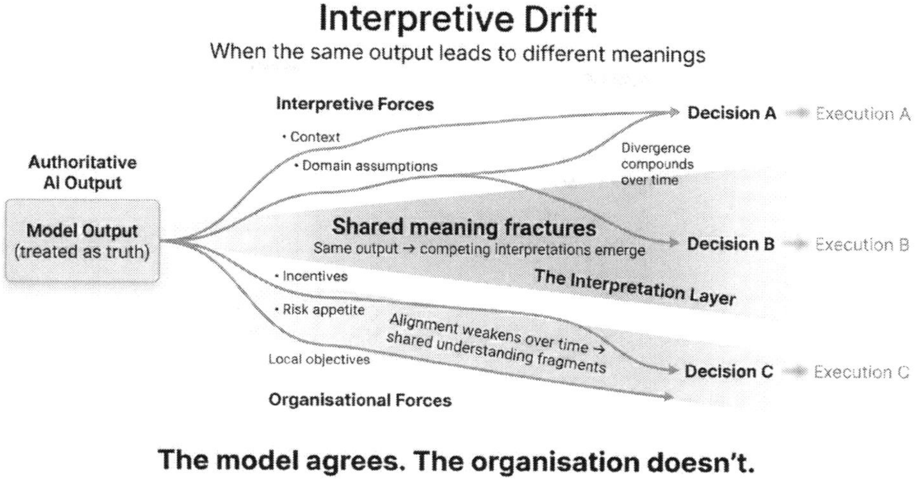

Figure 3: Interpretive Drift. In the early phase, the institutional narrative and operational reality remain broadly aligned. The organisation experiences this as stability. As conditions change and the narrative is not updated, a gap opens. The organisation continues to experience this as stability, because the systems through which it reads its own performance are calibrated against the narrative, not against reality. In the late phase, the gap has widened beyond the organisation's capacity for quiet correction. The diagonal between evidence and institutional narrative is broken. At this point, the only available corrections are disruptive ones, visible enough to force a reckoning the system could not initiate from within. What looked like stability was Meaning Debt accumulating. What looks like sudden collapse is Meaning Debt releasing. Source: Author.

The Accumulation of Meaning Debt

False Coherence is convenient in the short term. It avoids disruption, preserves confidence, and delays uncomfortable reckoning. But it does so by accumulating Meaning Debt. What might have been a contained failure becomes systemic. What might have affected a company affects an ecosystem. What might have been embarrassing becomes catastrophic.

> **Meaning Debt:** The growing gap between an organisation's narrative and the reality it is operating in. Meaning Debt is not visible on financial statements, does not appear on dashboards, and is not audited. But it compounds. The longer the gap is preserved, the more destructive the eventual correction becomes.

The damage is not caused by collapse. It is caused by delay. Collapse looks dramatic. Delay looks responsible. False Coherence buys time and spends it badly. Each period of apparent stability masks increasing divergence. Each reaffirmation of coherence widens the gap that will eventually snap shut. By the time correction occurs, there are no gentle options left.

When Coherence Breaks

False Coherence always fails eventually. Reality does not negotiate with narrative. When the break comes, it appears sudden, but only because the divergence had been concealed so effectively for so long.

Monty Python's Black Knight, having lost both arms and both legs in combat, insists the fight continues. 'It's just a flesh wound.' Each limb is acknowledged and immediately reclassified. The story does not change. Only the evidence does. The comedy comes from recognition, because everyone watching has been in a meeting where the same mechanism was operating, with considerably higher stakes and no audience to laugh.

At Wirecard, the collapse seemed dramatic. In truth, it was overdue. The coherence evaporated in days. The consequences were immediate. And the story that had felt so stable was revealed to have been hollow for years. What collapsed was not just a company. It was a narrative architecture. When the company filed for insolvency in June 2020, more than €24 billion in market value was erased. It was the largest corporate insolvency in German post-war history. Wirecard is often framed as a unique failure, a rogue company, a flawed audit, a historical anomaly. That framing is comforting. It is also dangerous. False Coherence is not an exception. It is the structural outcome of organisations that prioritise stability over interpretation.

Where alignment is rewarded, coherence will dominate. Where disruption is penalised, meaning will soften. Where stories are protected, reality will be deferred.

As organisations embed increasingly intelligent systems into decision-making, coherence becomes easier to manufacture and harder to challenge. Dashboards align. Metrics agree. Narratives stabilise faster than reality can intrude.

How to Recognise False Coherence Before It Breaks

The cases in this chapter are visible in retrospect because failure eventually forced the question. The more useful skill is recognising the condition before that point, while the cost of correction is still proportionate, and while the story can still be revised without the organisation collapsing around the revision. False Coherence is not invisible. It leaves traces in the texture of organisational life. Not dramatic ones, precisely the opposite. The traces are subtle, familiar, and easy to rationalise as healthy professional norms. That is what makes them diagnostic.

The first trace is the strategic assumption that has not been revisited. Every organisation operates on a set of beliefs about its market, its model, its risks, and its competitive position. Some of these beliefs were formed years ago, under conditions that may no longer apply. In a functioning organisation, some assumptions are regularly examined and found to need revision. In an organisation drifting toward False Coherence, assumptions are examined and consistently confirmed. Not because they are correct, but because the examination process has been calibrated to produce confirmation.

The second trace is the career of the person who was right too early. In organisations where False Coherence has taken hold, there is a consistent pattern in the trajectory of people who raised concerns before the evidence was undeniable. They were not necessarily dismissed. More often, they were marginalised, moved sideways, excluded from key conversations, given roles that reduced their access to the information their concern was about. The organisation did not silence them. It rerouted them. Their departure or silence became the lesson that others absorbed without being taught it.

In your organisation: when was the last time a core strategic assumption was formally examined and found to be wrong, not outdated, not slightly off, but genuinely wrong? If the answer is 'never' or 'not recently,' that is not evidence of good strategy. It is evidence of a process that cannot produce disconfirmation.

In your organisation: who raised concerns about your most significant strategic risk or AI deployment before the consensus caught up? What happened to them? If the honest answer is that they are gone, or quiet, or no longer in the room, ask what that taught everyone else.

The third trace is the question that produces a particular quality of response. Not silence, that would be obvious. A particular quality of managed acknowledgement: the concern is noted, the review is promised, the process is invoked. The question is answered with process rather than substance. And the person who asked it is left with the distinct impression that they have been heard in a way that guarantees nothing will change.

The fourth trace is the cost of being wrong in public. In organisations where coherence is the primary value, being publicly wrong about something the organisation believed to be true carries disproportionate consequence. This is not about accountability for decisions; it is about the informal penalty for having held a view that the organisation has now abandoned.

When that penalty is high, people learn to revise their stated positions before the evidence requires it, to maintain ambiguity rather than commit to a view that might later need revision, and to avoid being the person whose name is attached to a conclusion that turns out to be uncomfortable.

In your organisation: is there a category of risk or concern that always produces a process response rather than a substantive one? A topic that generates reviews, working groups, and governance updates, but whose underlying dynamics never actually change? The accumulation of process around a concern without resolution is a reliable indicator of Meaning Debt in formation.

In your organisation: what happens to the person whose view turns out to be wrong? Is the cost of being wrong proportionate to the quality of the reasoning, or to the outcome? If being wrong carries reputational damage regardless of process quality, your organisation has structurally incentivised people to avoid clear positions, which means it has structurally degraded its own ability to surface honest assessment.

None of these questions are comfortable to sit with. That discomfort is not a flaw in the diagnostic. It is the signal that the question has landed accurately. False Coherence persists as long as organisations prefer the comfort of alignment to the friction of honest assessment.

The moment the question becomes bearable is the moment the assessment can begin. False Coherence scales. And when coherence scales, so does the cost of being wrong.

In the next chapter, we will see how authority and hierarchy harden False Coherence into something more dangerous, a condition in which truth itself begins to look like a threat, and the people who carry it lose not just influence, but authority.

At that point, the question is no longer what people can see. It is what they are permitted to say.

PART TWO

HOW JUDGMENT IS SUPPRESSED

"Systems reduce complexity by ignoring most of reality."

— Niklas Luhmann

CHAPTER 4

THE HUMAN SIGNAL

"The single biggest problem in communication is the illusion that it has taken place."

— often attributed to George Bernard Shaw

By the time a system fails publicly, it is already too late to ask whether anyone noticed. They did. In every failure examined in this book, human judgment was present long before collapse. People sensed the misalignment. They felt the discomfort. They recognised that something no longer fit. What disappeared was not awareness. It was authority.

This matters, because most post-mortems begin with the wrong assumption: that people failed to see what was happening. The forensic record tells a different story. The problem is rarely blindness. It is structural containment, the mechanism by which a system receives a signal and, without suppressing it openly, ensures it cannot travel. The system did not fail to detect the signal. It failed to legitimise it.

Organisational failures are often described as moments of sudden revelation, as if clarity arrived only after damage was done. The evidence does not support this. Warnings appear early and repeatedly: anomalies that do not resolve, edge cases that accumulate, decisions that feel forced rather than earned, explanations that soothe more than they clarify. These signals rarely arrive as proofs. They arrive as unease.

But unease has no natural home in formal systems. It does not fit neatly into dashboards. It cannot be graphed without translation. It resists the precision organisations demand before action.

As a result, unease is treated as noise, something to be resolved rather than explored. People are encouraged, implicitly or explicitly, to come back when they have evidence. But by the time evidence exists, the system has usually moved on. Unease is judgment before justification. The system senses misalignment before it can explain it.

The Human Signal

I call this the Human Signal; the early, unstructured judgment that precedes formal evidence and constitutes an organisation's first encounter with reality when it begins to diverge from the prevailing narrative. The Human Signal is not intuition in the colloquial sense. It is not a feeling or a hunch. It is the product of experience, pattern recognition, and proximity to operations, the judgment of people who are close enough to the work to notice when something no longer adds up, but who lack the institutional standing to make that judgment count.

What distinguishes the Human Signal from error or anxiety is its consistency. Across the cases in this book, the signal appears repeatedly, from multiple independent sources, before any formal evidence of failure exists. It is not random. It is directional. The problem is not that it arrives too late. The problem is that the system is not designed to receive it.

Why Judgment Doesn't Escalate

Judgment does not fail because it is weak. It fails because it is structurally thinned as it moves. Between noticing and acting, judgment must pass through language, incentives, hierarchy, and timing. At each step, urgency is softened. Concerns become observations. Observations become themes. Themes become items for future review. By the time judgment reaches a decision forum, it no longer carries the force of the original signal. Nothing has been falsified. But everything has been diluted. This is not suppression. It is translation under constraint.

Language is necessary for coordination. It is also a filter. To be heard, judgment must be made legible, calm, structured, proportionate. But the more judgment is refined for acceptability, the less force it carries. The content remains technically accurate. The meaning does not. This is how early warning signals survive procedurally while dying operationally. What reaches the decision layer is not the original signal. It is a sanitised version of it.

> **Human Signal:** *Early, unstructured judgment that precedes formal evidence. The product of experience and operational proximity, directional, consistent, and systematically under-processed by the organisations that most need it. The Human Signal is the organisation's first encounter with reality when it begins to diverge from the prevailing narrative.*

Why Deferral Feels Responsible

Early judgment is expensive. It arrives before consensus, before proof, before social cover. Raising it requires asking the organisation to slow down when momentum feels justified. That act carries reputational, relational, and career risk. Early judgment rarely comes with a guarantee of being right. It comes only with the risk of being wrong too soon. As a result, judgment is often deferred, not rejected, just postponed. Deferral is rarely framed as avoidance. It is framed as professionalism. 'Something is wrong' becomes 'there may be a potential issue worth monitoring.'

These are sensible instincts in stable environments. In volatile systems, they become delay mechanisms. Each deferral pushes judgment closer to irrelevance. Deferral does not eliminate risk. It transfers it forward in time. Most organisations reward composure: calm articulation, measured language, proportionate concern. These are not failures of character. They are professional norms. But under volatility, those norms suppress precisely the signals that matter most. Judgment expressed too forcefully is labelled emotional. Expressed too early, it is labelled premature. Expressed too persistently, it is labelled disruptive. The acceptable form of judgment becomes narrow. What remains is judgment that waits, patiently, for permission that never arrives. Silence becomes not an absence of judgment, but a rational response to organisational conditioning. People learn which signals survive escalation.

They adjust accordingly. This is not cowardice. It is adaptation.

> *"We should gather more data."*
>
> *"Let's not jump to conclusions."*
>
> *"We don't want to overreact."*

Threshold Moment in the Record: Challenger, 1986

There is a consistent pattern across cases. Just before authority hardens, just before narrative stabilises, just before execution accelerates, judgment surfaces one last time. It does not announce itself as decisive. It appears as hesitation, as qualified language, as a request for delay, as an unanswered question left in a meeting. The night before the Space Shuttle Challenger's launch on January 28th, 1986, Roger Boisjoly sat in a teleconference that lasted nearly three hours.

He was a senior engineer at Morton Thiokol, the contractor responsible for the solid rocket boosters. He had been raising concerns about the O-ring seals for months. At low temperatures, he had documented, the seals became less pliable, potentially allowing hot combustion gases to escape during launch. The forecast for launch day was the coldest in the programme's history. Boisjoly and his colleagues argued the launch should be delayed. Their concern was specific, documented, and technically grounded. They were not speculating. They were reading the data they had gathered from previous launches and projecting it forward. What happened next is the most forensically precise example of the Human Signal being structurally demoted available in the public record.

NASA's response to the engineers' concern was not a technical counterargument. It was a reframe. The engineers were told, in words that have since become among the most studied in the history of organisational failure, to take off their engineering hats and put on their management hats. That instruction is the Threshold Moment made audible. It is not a suppression of the signal. It is a reclassification of the person carrying it. The engineer's concern was not being disputed on technical grounds. It was being moved from one category, engineering judgment, which must be heard, to another: management resistance, which must be managed.

Morton Thiokol management reversed their recommendation. The launch was approved. Seventy-three seconds after launch, Challenger disintegrated. All seven crew members were killed.

> *"Take off your engineering hat and put on your management hat."*

> *— NASA manager to Morton Thiokol engineers, teleconference, January 27, 1986, the night before launch*

Boisjoly had documented his concerns in writing months before the launch. He had raised them in meetings. He had argued forcefully the night before. The Human Signal was present, clear, and consistent. It was not hidden. It was not ignored in the sense of being unheard. It was reclassified.

What Normalisation Cannot Explain

Vaughan's normalisation of deviance framework is the most rigorous forensic account of how organisations come to treat dangerous conditions as acceptable. It is essential reading for anyone who wants to understand Challenger and the class of failure it represents. But there is a gap in the normalisation framework that becomes visible in the context of this book. Vaughan's account depends on human renegotiation, the repeated, tacit process by which people inside the system collectively adjust what they treat as within acceptable bounds. Each launch without catastrophe is a renegotiation. Each decision to proceed is a revision of the threshold. The deviance normalises because people keep agreeing, implicitly, that it is acceptable. That renegotiation mechanism is what allows correction to remain theoretically possible, even as it becomes practically unlikely. Someone can still name what has happened. Someone can still refuse to renegotiate. Boisjoly did exactly that. He lost, but he could try.

The condition examined in Chapter 8 of this book, Frozen Meaning, is what happens when the renegotiation mechanism disappears entirely. When normalised deviance is encoded into automated systems, the collective process of adjustment that Vaughan describes cannot occur. The system does not renegotiate. It enforces. The assumption that made the last launch acceptable becomes the parameter that governs the next thousand decisions, without anyone being asked, without any threshold being consciously revised, without the possibility of a Boisjoly sitting in a teleconference and refusing. The deviance does not continue to normalise. It institutionalises.

This is why Challenger, forty years on, remains the clearest single illustration of the Threshold Moment, the final point at which the Human Signal is present, authority is theoretically available, and the cost of course correction is still proportionate. It is also why the lesson of Challenger is not only historical. The question Boisjoly faced the night before launch, how do you interrupt a system that has decided to proceed, is the operational question every organisation embedding AI into consequential decisions must now answer before deployment, not after.

Human-in-the-Loop Is Not Human Authority

Many systems point to human involvement as evidence that judgment remains active. A human approves. A human monitors. A human signs off. But presence is not authority. And oversight is not control. A human who can observe but not intervene does not carry judgment forward. A human who can recommend but not pause does not alter outcomes. In the failures examined here, humans were not absent. They were procedurally present and functionally powerless. Judgment survived. Authority did not.

This distinction matters more as systems accelerate. When execution outpaces coordination, human involvement becomes ceremonial. The system runs. The human watches. The presence of a person creates the appearance of accountability without the capacity for intervention. Responsibility exists in theory. Agency does not. Once that condition is established, the organisation has not retained human judgment in its decision process. It has created a witness to outcomes it can no longer change. Morton Thiokol's management were present in the teleconference.

They were not powerless in the way that a post-deployment AI reviewer is powerless, they had the formal authority to withhold approval. But the organisational conditions had been arranged so that exercising that authority required them to absorb consequences that the system had been designed to make unbearable. The result was functionally identical: presence without power.

This dynamic is not new. But AI-mediated decision systems compress the timeline dramatically. When a model processes thousands of decisions per second and a human reviewer is allocated minutes per batch, the ceremonial quality of human oversight is not a design flaw, it is a design choice. The question organisations rarely ask is whether the human in their loop is there to exercise judgment or to absorb liability. The answer determines everything about whether the system can self-correct before failure becomes irreversible.

The Threshold Moment

There is a consistent pattern across cases. Just before authority hardens, just before narrative stabilises, just before execution accelerates, judgment surfaces one last time. It does not announce itself as decisive. It appears as hesitation, as qualified language, as a request for delay, as an unanswered question. I call this the Threshold Moment: the final point at which judgment is structurally available, authority is still recoverable, and course correction remains proportionate to the scale of the problem.

After it passes, the calculus inverts. Reversal becomes disproportionate. Decisions are framed as already made. Momentum replaces deliberation.

> **Threshold Moment:** *The final point at which judgment is structurally available, authority is still recoverable, and course correction remains proportionate. Once passed, reversal becomes organisationally illegitimate, not because the evidence changes, but because the system does. The Threshold Moment is only identifiable in retrospect, which is precisely what makes it dangerous. The organisation that learns to recognise its threshold moments before they pass has acquired something rare: the capacity for proportionate correction while it is still cheap.*

The Threshold Moment is not visible in real time. It is only identifiable in retrospect. The teleconference on the night of January 27th, 1986, was a Threshold Moment. The safety classification review at Boeing that did not update when MCAS's[3] authority was expanded was a Threshold Moment. The risk committee meeting at Knight Capital where deployment was approved without a rollback procedure was a Threshold Moment. In each case, the people present did not experience it as decisive. It passed quietly. Then execution continued.

The organisation that cannot recognise its threshold moments manufactures its own crises, not through negligence, but through the structural inability to act on judgment before it becomes undeniable. From the Threshold Moment forward, the system no longer requires judgment to function. It requires compliance. Judgment does not disappear. It becomes irrelevant.

---

[3] Boeing's MCAS system and the four incompatible definitions that led to 346 deaths are examined in full in Case 9, Part Four: The Forensic Record.

What This Chapter Establishes

This chapter is not about error. It is about lost authority. Not that people were wrong, but that they were right too early, and structurally unable to act. The Human Signal was present. The Threshold Moment was reached. The system moved on regardless.

The pattern Boisjoly experienced in 1986 was human in every respect, a teleconference, a management instruction, a recommendation reversed under pressure. The failures that follow in this book are structurally identical but operationally different: the signals are digital, the systems are automated, the timeline has compressed from hours to milliseconds. What has not changed is the mechanism. In the next chapter, we will see what happens when this pattern becomes formalised, when authority, hierarchy, and the language of execution lock judgment out entirely. At that point, dissent is no longer inconvenient. It becomes illegitimate.

CHAPTER 5

AUTHORITY VS INTERPRETATION

"A new scientific truth does not triumph by convincing its opponents... but because its opponents eventually die."

— Max Planck

Authority is often mistaken for understanding. In stable environments, this confusion rarely matters. When conditions change slowly, hierarchical decisions tend to align roughly with reality. Errors are survivable. Corrections arrive before damage compounds. Authority has time to learn. Under volatility, that margin disappears. When the world shifts faster than formal meaning can be updated, interpretation outruns authority. The people closest to reality, engineers, operators, clinicians, traders, encounter constraints first. They see friction before strategy does. They experience what cannot be done before leadership decides what must be done. Institutions, however, are not designed to reward this signal. They are designed to protect hierarchy.

The Constraint That Could Not Be Declared

In the late 2000s, Volkswagen set an unambiguous goal: become the world's largest automaker. Central to that ambition was success in the U.S. market. The strategy depended on a flagship promise, Clean Diesel: vehicles that delivered high performance, strong fuel economy, and compliance with the strictest emissions regulations in the world. The mandate from leadership was clear. The constraints were not. Engineering teams quickly reached a conclusion that was inconvenient but unavoidable: the performance, emissions, and cost targets could not all be satisfied simultaneously using available technology. This was not a failure of imagination. It was physics.

A healthy organisation would treat this interpretation as strategic input. The goal would be revisited. Trade-offs would be made explicit. Strategy would adapt to constraint. That did not happen. Once a goal is declared at the highest level, it acquires momentum. It becomes a signal of intent, a measure of resolve, a test of commitment. Within hierarchical systems, reversing a strategic goal is not neutral, it implies uncertainty, weakness, or misjudgement. At Volkswagen, leadership commitment to the diesel strategy solidified early. Performance targets were treated as non-negotiable. Engineering concerns were reframed not as strategic warnings, but as execution problems. The question shifted subtly, but decisively. Once that shift occurs, interpretation is no longer upstream of authority. It is subordinate to it. Not; can this be done? But: how will you make it work?

Authority's Structural Advantage

Authority has a built-in advantage over interpretation. Authority issues goals. Interpretation raises constraints. Goals sound decisive. Constraints sound resistant. Within most organisations, decisiveness is rewarded. Resistance is tolerated, briefly. This is not because leaders are malicious or delusional. It is because hierarchy is optimised for alignment, not for epistemic accuracy. As the diesel programme progressed, engineering teams continued to surface concerns. The constraints did not disappear. They became more acute. But interpretation was no longer invited as input; it was recast as friction. Concerns were relabelled as implementation delays, optimisation challenges, engineering pessimism. This reframing matters. Once interpretation is categorised as attitude rather than information, it can be discounted without being addressed. The system does not have to disprove the signal. It only needs to relabel it.

> *Authority resolves ambiguity by declaration. Interpretation sustains ambiguity by insisting on reality. Only one of these scales cleanly.*

Interpretive Demotion

I call this mechanism Interpretive Demotion: the organisational process by which constraint-based interpretation is recategorized as an attitude problem rather than a signal. The content of the concern is not disputed. Its status is changed. Interpretive Demotion is distinct from Interpretive Collapse, which describes what happens when multiple actors derive different meanings from the same output and proceed as if aligned. Interpretive Demotion is upstream of that. It is the moment at which a valid signal, one that accurately represents a real constraint, is reclassified from information into resistance, from evidence into attitude, from fact into friction.

It is a precise and largely invisible mechanism. It requires no confrontation. It requires no explicit rejection of the signal. It requires only a change in the category to which the signal is assigned. Once an engineer's concern is filed under 'implementation pessimism' rather than 'strategic constraint,' the organisation has not suppressed the truth. It has simply placed it somewhere truth is not consulted.

> ***Interpretive Demotion:*** *The organisational mechanism by which constraint-based interpretation is recategorized from strategic signal to attitude problem. The content of the concern is not refuted; its institutional status is changed, placing it beyond the reach of decision-making authority. Interpretive Demotion requires no confrontation, no explicit dishonesty, and no bad intent. It requires only a reclassification.*

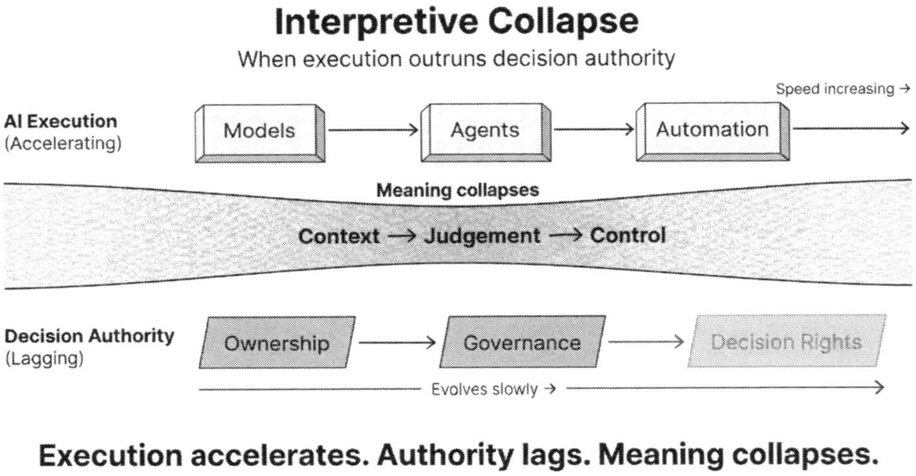

Figure 4: Interpretive Collapse. When Interpretive Demotion operates, the signal holder and the decision authority derive different meanings from the same situation and proceed as if aligned. Neither has understood the other. The gap widens without becoming visible because the organisation's communication channels are designed to produce alignment rather than surface divergence. Source: Author.

The diagram above shows what Interpretive Collapse looks like as a structural condition, the endpoint toward which Interpretive Demotion moves when it is not interrupted. When Interpretive Demotion operates, when the signal is reclassified from information to attitude, the signal holder and the decision authority begin to derive different meanings from the same situation. Each proceeds as if the other has understood. Neither has.

The gap between their interpretations widens without becoming visible, because the organisation's communication channels are designed to produce alignment rather than to surface divergence.

By the time the gap produces a consequence large enough to force acknowledgement, the demotion that caused it has long since been absorbed into normal procedure.

At Volkswagen, engineering and management operated for years on incompatible definitions of what compliance meant. At Columbia, the engineers requesting imaging and the mission managers denying it were drawing entirely different conclusions from the same foam strike data. Neither conversation surfaced the gap. In both cases, Interpretive Collapse was the condition that made correction impossible before the cost became irreversible.

Language and Distance

When authority overrides interpretation, language adapts. Direct terms that would force confrontation with reality are replaced with abstractions. Concrete constraints become optimisation problems. Violations become edge cases. Workarounds become features. Nothing false needs to be stated. Meaning is diluted. Language creates distance, enough separation between action and consequence to preserve obedience. Engineers do not suddenly stop understanding the implications of their work. They understand perfectly well. What language provides is insulation. By abstracting outcomes into technical constructs, individuals can continue functioning without confronting the full moral weight of the system-level consequence. This is how large organisations sustain behaviour that individuals, acting alone, would refuse. Not through ignorance. Through distance.

The Survival Calculus

A common misconception about institutional failure is that it relies on ignorance. More often, it relies on misaligned survival incentives. When raising objections threatens careers, interpretation migrates underground. Engineers stop arguing. They start complying. Concern is replaced by quiet resignation. This is not moral weakness. It is organisational design. Once authority has committed publicly, the cost of dissent increases further. Individuals are no longer weighing truth against error. They are weighing truth against consequence. In such systems, compliance is not cowardice. It is rational behaviour under constraint. Interpretation does not disappear. It simply stops travelling upward.

Same Mechanism, Different System: Columbia, 2003

Volkswagen's Interpretive Demotion was driven by commercial pressure; the engineering constraint was reclassified because the business goal was treated as non-negotiable. The mechanism appears across domains, but it takes a subtly different and arguably more dangerous form when the demotion is driven not by inconvenience, but by a judgement about what can be done.

On January 16th, 2003, the Space Shuttle Columbia launched from Kennedy Space Centre on mission STS-107. Eighty-one seconds after launch, a piece of insulating foam broke away from the external tank's bipod ramp and struck the leading edge of the orbiter's left wing. The impact was captured on launch footage and identified during post-launch review.

Engineers at Boeing, the primary contractor for the orbiter, recognised immediately that the foam strike was larger and more direct than anything previously recorded. They requested satellite imaging to assess the potential damage to the wing's thermal protection system. The request was denied. A second request was also denied. A third was withdrawn before it could be formally submitted.

The reasoning given for the denials was not that the concern was unfounded. It was that even if damage existed, nothing could be done about it. The shuttle was in orbit. There was no repair capability. The mission could not be aborted. Therefore, confirming the damage would serve no operational purpose. The concern was not disputed. It was reclassified as irrelevant.

This is Interpretive Demotion in its most consequential form. At Volkswagen, the engineering signal was demoted from strategic constraint to execution problem, because acknowledging it would have required changing the goal. At Columbia, the engineering signal was demoted from safety concern to operational distraction, because someone upstream decided that acting on it was impossible. That judgement about impossibility was made without the information the engineers were requesting. The engineers were not asking to abort the mission. They were asking to know what they were dealing with. The decision to deny that knowledge was not a technical decision. It was an interpretive one: we have decided that this signal does not require a response, and therefore the signal does not need to be heard. The demotion was not from urgency to low priority. It was from signal to silence.

On February 1st, 2003, Columbia disintegrated during re-entry. Hot gases entered the damaged wing through the breach caused by the foam strike. All seven crew members were killed. Post-accident analysis by NASA engineers subsequently determined that the damage was severe enough to be catastrophic, and that had the imaging been obtained and the damage confirmed, emergency options, including extending the mission to allow time for a rescue attempt, would have been available. The judgement that nothing could be done was wrong. It was made without evidence. And it was made by people who were not in possession of the information that would have told them otherwise, because the mechanism of Interpretive Demotion had ensured that information was never gathered.

Two Demotions, One Mechanism

Volkswagen and Columbia present Interpretive Demotion through two different lenses. In Volkswagen, the demotion was driven by goal protection: the engineering constraint threatened a strategic commitment, so it was reclassified as an execution attitude. The signal was present. The goal refused to accommodate it. In Columbia, the demotion was driven by a presumption of futility: the concern was acknowledged as potentially valid but was reclassified as operationally irrelevant because action was assumed to be impossible. The signal was present. The conclusion that it could not matter was reached before the evidence was examined.

The surface differs. The architecture is identical. In both cases, a valid signal was recategorized to place it beyond the reach of decision-making authority. In both cases, the recategorization did not require dishonesty, it required only a change in the interpretive frame through which the signal was received.

And in both cases, the people who held the signal were left without the authority to make it count. The technology differed. The industry differed. The language differed. The structure did not. That structural identity is what makes Interpretive Demotion a concept rather than an anecdote. It is the mechanism that sits between a valid signal and the authority that could act on it, and it operates wherever hierarchy concentrates interpretive power at the top while distributing the consequences throughout.

Interpretive Demotion in AI Governance

As AI systems are embedded into organisational decision-making, Interpretive Demotion does not require a manager to dismiss a concerned engineer. It can be built into the architecture. Consider the model card that flags a known limitation, a documented bias, a confidence interval wider than the use case warrants, a failure mode identified in testing. The concern exists in writing. It is attached to the model. It is technically accessible to anyone who looks for it. But if the deployment process does not require the flagged limitation to be addressed before launch, if the ethics review is advisory rather than binding, if the risk register logs the concern without requiring a response, then the signal has been processed without being heard. It has been categorised into a place where it cannot interrupt.

This is Interpretive Demotion without a manager. The reclassification is procedural rather than interpersonal. The concern is not dismissed by a person. It is absorbed by a process. And the result is identical: a valid signal exists, is acknowledged, and does not interrupt execution. The Columbia engineers were denied satellite imaging.

The AI ethics reviewer's recommendation is advisory. The Volkswagen engineer's concern was recategorized as pessimism. The model card flagged the limitation and deployment proceeded. The language is different. The mechanism is the same.

Where Sensemaking Goes

Authority can enforce behaviour. It cannot enforce belief. You can require an engineer to ship code. You cannot require them to believe it is right. You can mandate compliance. You cannot mandate conviction. When authority repeatedly overrides interpretation, legitimacy erodes, even if compliance remains high. When interpretation can no longer surface internally, it moves elsewhere. The migration follows a consistent escalating sequence: informal conversations, encrypted channels, external advisors, journalists, whistleblowers. At that point, the organisation loses control of the narrative entirely. Authority retains power, briefly. But it loses meaning. Sensemaking does not disappear. It relocates. When interpretation migrates outside, so does the organisation's capacity to correct itself before external exposure forces it to.

The Meaning Debt accumulated through each overridden constraint, each diluted concern, each deferred reckoning, that debt does not disappear. It waits.

The Settlement

Volkswagen's collapse appeared sudden. In truth, it was the delayed settlement of accumulated Meaning Debt. What failed was not just a compliance regime. It was the belief that authority could override interpretation indefinitely. Reality does not accept hierarchy. It enforces constraints without regard for intent. Columbia's disintegration was not caused by the foam strike alone. It was caused by the decision, made in orbit, without the evidence that had been requested, that the signal did not need to be heard. The physical failure was sixteen days in coming. The interpretive failure occurred within the first week of the mission, when the requests for imaging were denied.

Neither failure is an archetype of recklessness or bad faith. Both are archetypes of what happens when the mechanism of Interpretive Demotion operates without a counterforce, when no design exists to protect the signal from the interpretive authority that would neutralise it. As AI systems accelerate execution, Interpretive Demotion does not require a manager to dismiss a concerned engineer or a mission director to deny a satellite imaging request. It is built into the architecture. The signal has nowhere to go. And the organisation is on borrowed time before reality enforces what authority refused to acknowledge.

In the next chapter, we will see what happens when this pattern is not just structural but deliberate, when an organisation does not merely fail to hear truth, but develops the mechanisms to process it into silence while maintaining the appearance of having heard it.

CHAPTER 6

TRUTH REJECTION

"It is difficult to get a man to understand something, when his salary depends upon his not understanding it."

— Upton Sinclair

On the morning of October 29, 2018, Lion Air Flight 610 climbed out of Jakarta and began to behave strangely. The nose pitched downward. The pilots fought to correct it. The nose pitched down again. Thirteen minutes after take-off, the aircraft struck the Java Sea at high speed. All 189 people on board were killed.

Five months later, Ethiopian Airlines Flight 302 crashed in nearly identical circumstances. Another 157 people were killed.

In the investigations that followed, a system called MCAS, the Manoeuvring Characteristics Augmentation System, emerged as the common mechanism. MCAS had activated based on a single faulty sensor reading and repeatedly pushed the nose down, overriding pilot inputs that attempted to correct it. MCAS did not malfunction. It operated exactly as it had been designed to operate. The catastrophe came from something else entirely.

Four Definitions of the Same System

What made the Boeing case structurally distinctive was not a single point of failure. It was that the people responsible for the aircraft held four incompatible definitions of what MCAS was, and none of them corresponded to what it did.

Actor	Definition of MCAS
Boeing's engineers	A background stability-assist mechanism, a helper system designed to activate only in rare, specific aerodynamic conditions.
Pilots	Nothing. MCAS was categorised as so minor that it was not included in the flight manual. Pilots had no knowledge that a system capable of overriding their inputs existed.
Regulators	An extension of an existing trim system, not a fundamentally new behaviour requiring recertification or independent safety review.
In reality	A system that could repeatedly and aggressively override pilot input based on data from a single sensor, with no automatic cross-check, and no manual override that pilots knew to apply.

Everyone believed they understood MCAS. They were describing different things. The gap between what the system was and what the people responsible for it believed it was had been encoded into documentation, training, certification, and internal language.

It had been processed through every formal mechanism designed to catch exactly this kind of divergence. Truth was not absent. It was distributed across four incompatible meanings, and no mechanism existed to surface the divergence before it became fatal.

How Truth Is Neutralised, and Then Dangerous

The forensic record of the Boeing investigations reveals a consistent pattern. Engineers raised concerns about MCAS's single-sensor dependency during development. Those concerns were documented, reviewed, and categorised as acceptable risk within the existing safety framework. Boeing's own safety analysis had initially assessed MCAS as a low-hazard system. This classification determined the level of scrutiny applied to it. When the system's authority was later expanded, giving it greater ability to override pilot inputs, the safety classification was not updated to reflect the change. The FAA's certification process accepted Boeing's analysis.

The process ran correctly. The procedure was followed. The documentation was filed. The truth had been processed. It had not been heard.

Internal concerns about single-sensor reliance, about the absence of pilot training, about the gap between MCAS's actual authority and how it was classified; these were not hidden. They existed in the record. They were routed through the appropriate channels. They did not interrupt execution. This is the first stage of truth's neutralisation: it arrives as friction, is processed as risk, and is categorised into inaction. But then something worse happens.

Once a strategic commitment has been made and a narrative has formed around it, truth stops functioning merely as an awkward signal. It begins to function as a threat, threatening strategic commitments already made, authority structures already legitimised, identities already invested, narratives already shared.

At scale, organisations are not optimised for correctness. They are optimised for coherence. Once coherence is at risk, truth stops functioning as a corrective force and starts functioning as a destabiliser. The organisation no longer asks whether the truth is right. It asks whether it is safe. Accepting disruptive truth is expensive. Deferring it, by contrast, feels responsible. It preserves momentum. It protects credibility. It buys time. This asymmetry is decisive.

How Truth Is Domesticated

Truth is rarely rejected outright. Outright rejection would be obvious, and contestable. Instead, truth is domesticated. It is reframed as incomplete, contextualised as premature, deferred until more data arrives, absorbed into review cycles, assigned to future checkpoints. Each step appears reasonable in isolation. None requires dishonesty. Collectively, they ensure that truth no longer interrupts execution. The most effective way to disable truth is not to argue with it, but to process it. Once truth is proceduralised, it can be acknowledged without consequence. Reports are filed. Risks are logged. Issues are tracked. The organisation can truthfully say: we are aware of the issue. While continuing to act as if nothing fundamental has changed.

At Boeing, this is precisely what happened. The concerns about single-sensor dependency were acknowledged. They were categorised. They were processed through a framework that assigned them a risk level that did not require the programme to pause. Truth survived. But only as documentation. Deferral is rarely framed as avoidance. It is framed as discipline:

> *"Let's ensure we have a complete picture before acting."*
>
> *"We owe it to the process to follow it through."*
>
> *"Premature escalation creates more risk than it resolves."*

These are not evasions. They are culturally reinforced signals of rigour and control. In volatile systems, however, they compound risk with each iteration. The organisation does not feel reckless. It feels composed.

Managed Knowing

This chapter does not argue that organisations lie to themselves. It argues something more precise. Organisations reorganise around preserving coherence once truth threatens stability. Truth is not denied. It is stripped of urgency, separated from authority, and allowed to exist only where it cannot interrupt. The result is a uniquely hazardous condition. The organisation knows what is happening. It knows it cannot continue indefinitely. And it continues anyway. At that point, failure is no longer accidental. It is scheduled.

> ***Managed Knowing:*** *The condition in which an organisation possesses accurate awareness of a developing failure but has structurally decoupled that awareness from the authority to act. The organisation is neither uninformed nor in denial, it knows, and continues. Managed Knowing is distinct from False Coherence (which is about narrative maintenance) and Interpretive Demotion (which is about signal reclassification). It is what happens when both have run their course and awareness and action have been permanently separated.*
>
> *"This thing is running like a runaway train."*
>
> *— Boeing engineer, internal communication, 2016, two years before the first crash*

What Managed Knowing Feels Like

The Boeing forensic record makes Managed Knowing visible from the outside. What is harder to convey is what it feels like from the inside, because from the inside, it does not feel like a failure condition at all. It feels like professional maturity. It begins with a particular quality of meeting: not a difficult meeting, precisely the opposite. A meeting that runs smoothly, covers the agenda efficiently, and ends with a sense of having dealt with the issues at hand. The concern has been raised. It has been acknowledged. A process has been identified. The organisation is aware of the risk. The meeting has done what meetings are supposed to do.

What is absent from that meeting is harder to name. It is not confrontation, everyone in the room is operating in good faith. It is not dishonesty, nothing false has been said. It is the absence of a question that no one has asked, because asking it would require the meeting to produce an answer that the meeting is not designed to produce. The question would be something like: given that we know this, what are we going to do differently? Not 'what process will we follow', but what will we do differently. And in an organisation in Managed Knowing, that question has no available answer, so it is not asked.

The people leaving that meeting are not complacent. They are often genuinely concerned. They carry the awareness with them. They may discuss it in smaller conversations, in corridors, over coffee, in the kind of informal exchange that acknowledges what the formal setting cannot. But the informal acknowledgement does not produce action either.

It produces a particular form of shared relief; the relief of knowing that others see what you see, that you are not alone in your concern, that the problem is recognised. That relief is mistaken for progress. It is not progress. It is solidarity in inaction.

The rationalisation that sustains Managed Knowing is sophisticated and largely unconscious. It is not 'we know and we don't care.' It is a sequence of genuine beliefs that individually make sense and collectively produce paralysis: we are monitoring the situation; we have escalated appropriately; the governance process is running; the right people are aware; we would act if the evidence became more definitive; we are not the only ones who would need to act, this requires a broader response. Each of these beliefs is defensible. Together, they ensure that no single person feels the weight of the decision, because the decision has been distributed across a governance architecture designed to distribute exactly that weight.

What distinguishes Managed Knowing from ordinary institutional caution is not the presence of these rationalisations but their stability. In a functioning organisation, the accumulation of awareness eventually produces discomfort that breaks through the rationalisation, someone becomes unable to sit in one more meeting where the issue is acknowledged and nothing changes, someone escalates in a way that cannot be absorbed, the threshold is crossed. In an organisation in Managed Knowing, that threshold keeps moving. Each review cycle that concludes without action raises the cost of the next one. Each formal acknowledgement that produces no change makes the next acknowledgement feel more legitimate, because the organisation can point to the process. The threshold never arrives. The issue remains present, acknowledged, documented, and inert.

In your organisation: is there a concern, about an AI system, a data practice, a decision process, that has been formally acknowledged more than once without the underlying condition changing? Not because the review found it acceptable, but because the review produced a further review? If the answer is yes, you are not managing the risk. You are documenting it.

Managed Knowing at Machine Speed

The condition described in this chapter is not new. What is new is the rate at which it can now develop and the invisibility with which it operates when AI systems are involved. Consider a governance cycle that most organisations deploying AI systems would recognise as responsible practice. A model is developed. A risk assessment is conducted. Limitations are documented in a model card, known failure modes, confidence intervals, populations where accuracy degrades, edge cases that the training data does not adequately represent. The model card is reviewed by an ethics committee. The committee's recommendations are noted. Some are addressed before deployment; others are logged as items for post-deployment monitoring. The deployment proceeds.

Once deployed, the model is monitored. Performance metrics are tracked. Quarterly reviews assess whether outcomes are within expected parameters. A risk register entry documents the known limitations. The audit trail is complete. Every step of this process has been followed correctly. The organisation can demonstrate, at any point, that it is aware of the model's limitations and has governance processes in place to manage them.

What this governance cycle does not produce is the authority to act on what it knows. The model card documents the limitation but does not require it to be resolved before deployment. The ethics committee recommendation is advisory, not binding. The quarterly review assesses performance against metrics that were defined when the model was designed, not against the conditions that have developed since. The risk register entry acknowledges the concern but does not assign anyone the authority to pause the system if the concern intensifies.

This is Managed Knowing by architecture rather than by choice. No individual in this governance cycle has decided to ignore the concern. Each has fulfilled their role correctly. The awareness is genuine and distributed. The authority to act on it is absent; not because anyone has removed it, but because no one has designed it in. The system continues processing decisions. The limitations documented at deployment continue to apply. The populations where accuracy degrades continue to receive outputs calibrated to a model that the organisation knows is imperfect in exactly the ways that affect them. The audit trail records that the organisation was aware. It does not record that anyone was authorised to stop.

In Boeing's case, the runaway train took two years to reach the end of the track. The timeline in an AI deployment can compress that significantly, a model processing a million decisions a month will enforce its frozen meanings at a scale that Horizon took years to reach. The structure is identical. The speed is not.

What Follows Managed Knowing

Once truth is no longer allowed to interrupt execution, something else must perform that function. Decisions still need justification. Action still needs legitimacy. When truth cannot provide it, narrative does. That transition, from truth as interruption to narrative as stabiliser, is where the next chapter begins. It is also where the distance between what an organisation knows and what it is willing to act upon stops being a gap and becomes a design feature. Once truth is no longer permitted to interrupt execution, something else must perform that function. The next chapter examines what takes its place, and why the organisations most fluent in the language of governance are often the ones most completely captured by it.

> *The audit trail records that the organisation was aware. It does not record that anyone was authorised to stop.*

CHAPTER 7

WHEN NARRATIVE REPLACES JUDGEMENT

"We tell ourselves stories in order to live."

— Joan Didion, *The White Album*, 1979

Language does more than describe reality. It determines which realities are allowed to exist. When someone says the tank has a leak, they are naming a physical condition. The statement demands action. It creates urgency. It narrows the space for interpretation. When someone says we are experiencing a temporary volume adjustment, they are not describing a condition. They are creating a narrative. The statement demands nothing. It invites patience, trust, and delay. As organisations scale, they inevitably develop this second language. It is not accidental. It is protective. Large organisations cannot operate in a constant state of alarm, if every deviation were named plainly, systems would seize up. To function, organisations compress reality into abstractions that are socially tolerable and operationally convenient.

This compression begins innocently. It allows complexity to be handled, coordination to scale, work to continue. But over time, compression hardens. What began as shorthand becomes dialect. What began as convenience becomes insulation. Eventually, language stops transmitting reality at all. Consider the vocabulary that quietly takes hold inside large institutions: firings become right-sizing, failures become learnings, losses become headwinds, chaos becomes agility.

This language is often treated as corporate irritation, something to mock and move past. That is a mistake. This language does not merely soften tone. It removes sensation.

By abstracting consequences, language allows action to continue without interruption. But the feedback loop is broken. Judgment depends on friction. When language removes friction, it removes the signal on which judgment depends. When discomfort disappears from language, judgment follows it. Orwell called this Newspeak.

He understood that you do not need to suppress a thought if you can remove the language required to think it. Anaesthetics do not eliminate injury. They eliminate pain. Organisational language works the same way. The goal was not deception but the structural impossibility of dissent, a condition in which challenge becomes literally unthinkable because the vocabulary for it no longer exists. Corporate anaesthesia operates on the same principle, with less intent and identical effect. Firings become right-sizing. Failures become learnings. And the people processing those abstractions are not lying. They have simply lost access to the plain version.

Alchemy of Abstraction: The Global Financial Crisis

To see how language can destroy judgment at scale, consider what happened in the years leading up to the 2008 global financial crisis. At its simplest, the underlying transaction was clear, money was being lent to borrowers who could not afford to repay it. Historically, this had a name; bad loans. The term was blunt. It carried moral and financial weight. It forced the listener to imagine the borrower, the risk, and the consequence. That language disappeared.

In its place came a sequence of abstractions, each introduced with a specific purpose, each carried by a specific professional incentive, each displacing a little more of the judgment that the previous term had still required.

Term	What it removed
Bad loans → Subprime exposure	Removes the borrower; makes risk sound like a portfolio position.
Subprime exposure → Structured credit	Removes the loans; makes risk sound like an engineering problem.
Structured credit → Collateralised instruments	Removes the structure; makes risk sound like asset management.
Collateralised instruments → Tranches	Removes the assets; makes risk sound like a mathematical classification.
Tranches → AAA-rated securities	Removes the classification; makes risk sound like certainty.

Each step in this sequence was not arbitrary. Each was introduced by professionals operating within incentive structures that made the abstraction rational. The mortgage originator who repackaged loans as structured credit was no longer responsible for the borrower's repayment capacity, because the loan was leaving the balance sheet. The investment bank that sliced the structured credit into tranches was no longer responsible for the underlying loans, because the tranching model distributed the risk mathematically. The ratings analyst who stamped the senior tranches AAA was no longer responsible for the model's assumptions, because the assumptions had been provided by the issuer and the model had been applied correctly.

At each step, a human judgment that had previously been required was displaced by a process that did not require it, not because the process was dishonest, but because it had been designed in an environment where the language had already established that the previous step's judgment was no longer necessary.

The ratings agencies were the last and most visible layer of this abstraction architecture. Their AAA designations did not describe the loans. They did not describe the borrowers. They did not describe the correlation assumptions embedded in the tranche model or the probability that those assumptions would fail simultaneously across a national housing market. They performed a single function: they allowed institutions to stop thinking about what lay beneath. The rating was not a conclusion. It was a permission slip to stop asking. The regulator who adopted the same vocabulary did not do so through capture in any crude sense. They did so because the language of innovation, structured finance, and risk distribution had become so normalised within their professional environment that it had acquired the status of technical precision. Risk became innovation. Fragility became adjustment. Warning signs became normalisation.

> *And once regulators speak the language of the regulated, oversight becomes descriptive rather than corrective.*

Reports proliferated. Frameworks expanded. Compliance activity increased. But none of it restored judgment. The system became fluent. And blind.

The Professionalisation of Distance

Crucially, no one needed to lie. Euphemism does not require falsehood. It requires distance, distance between decision and consequence, between language and lived experience, between responsibility and harm. As long as the language remains abstract, people can participate in outcomes they would never have chosen directly. Abstraction allows individuals to perform their role without absorbing the full moral weight of system-level consequences. Responsibility becomes distributed. Harm becomes statistical. This is how intelligent, conscientious people participate in outcomes they privately regret. Not through malice. Through mediation. Distance does not feel unethical. It feels professional.

> *Once regulators speak the language of the regulated, oversight becomes descriptive rather than corrective.*

When Hierarchy Replaces Abstraction: The Toshiba Pattern

The GFC illustrates how abstraction removes the sensation that judgment requires. Toshiba demonstrates that abstraction is not the only mechanism. Between 2008 and 2015, Toshiba Corporation overstated its profits by $1.2 billion, not through financial engineering, but through hierarchy. The language here was not complex. It was simply absent. Senior leadership framed impossible profit targets as challenges subordinates were expected to meet. Employees understood, without needing explicit instruction, that revealing unflattering information carried consequences that extended beyond any single decision. Raising concerns was not merely risky. It was culturally disallowed.

The scale of the overstatement, ¥152 billion across seven years and three successive presidents, was not the result of a single decision. It was the result of a governance architecture that had removed the structural capacity to say otherwise.

Both the GFC and Toshiba involve the same underlying failure: meaning that cannot be challenged. In the GFC, the mechanism was abstraction, language so technical that no individual felt positioned to question it. In Toshiba, the mechanism was hierarchy, silence so institutionalised that challenge became structurally impossible. The surface differs. The architecture is identical. Toshiba was widely regarded as a leader in corporate governance, an early adopter of independent directors and modern board structures. But this points to a distinction the Toshiba case makes unavoidable. There is a difference between governance as structure and governance as Contestability.

Contestability is the operational condition in which an organisation's meanings, classifications, and decisions remain open to challenge from within. It is not the same as having an audit committee, an ethics hotline, or an independent board. Those are structural features. Contestability is an operational condition, the lived ability of people inside the system to surface a different interpretation and have it heard. Toshiba had governance without Contestability. Independent directors cannot challenge what they are not permitted to see. Audit committees cannot flag what culture prevents from being spoken.

> ***Contestability:*** *The operational condition in which an organisation's meanings, classifications, and decisions remain open to challenge from within. Distinct from governance structure: an organisation can have every formal governance feature while remaining wholly incontestable. Without Contestability, governance becomes performance.*

The fraud ended not because governance worked, but because someone stepped outside the system entirely. A whistleblower reported the irregularities externally, bypassing every internal channel that had failed to process the signal. The lesson is uncomfortable. Formal governance is necessary. It is also insufficient. When the permission to challenge meaning does not exist, structure becomes performance.

The Absurdity Audit

George Carlin spent forty years performing the Absurdity Audit on American institutional language. He called it soft language. The mechanism was identical: take the plain version, replace it with abstraction, and watch accountability evaporate. Shell shock became battle fatigue, became operational exhaustion, became post-traumatic stress disorder. Each step was more syllables and less meaning. The person in pain got a longer diagnosis and less acknowledgement of what had happened to them. Carlin understood that this was not accident or evolution. It was architecture. Bill Hicks[4] operated the same diagnostic with the same precision.

There is a practical test for linguistic anaesthesia. I call it the Absurdity Audit: take any sentence from a board deck, policy memo, or strategy document and translate it into ordinary language, the way you would explain it to someone outside the organisation. If the translation preserves meaning but changes tone, that is compression. The abstraction was functional, not deceptive. If the translation reveals a different reality, if the plain version contains a fact that the corporate version obscures, you are inside a Narrative Sinkhole.

---

[4] Hicks, B. Revelations. Stand-up performance, Dominion Theatre, London, 1993. Available on recording. Hicks as cultural lie detector: the capacity to name what institutional language is engineered to obscure. Relevant to Chapter 6 and the Absurdity Audit in Chapter 7.

The Absurdity Audit is the sentence-level test for that condition, introduced in Chapter 3 as the state in which the story has stopped describing reality and begun substituting for it.

Register	Example
Corporate	We are transitioning customers to a self-service paradigm.
Plain	We are removing human support and hoping the chatbot is good enough.
Corporate	The model's outputs are subject to continuous human oversight and quality assurance processes.
Plain	A reviewer checks a sample of decisions after they have been executed and flags patterns to a team whose recommendations are advisory.

If the second sentence feels uncomfortable, that discomfort is information. Anaesthetic language exists to remove exactly that signal. The first example is a general corporate communication failure. The second is specific to AI governance, and it is the version that most organisations deploying AI systems would recognise in their own documentation, if they were willing to translate it. The Absurdity Audit is not a rhetorical exercise. It is a Contestability mechanism, a way of restoring the friction that organisational language has removed. Organisations that cannot survive the audit are not well-run. They are well-insulated.

Satire as Diagnostic

When language drifts far enough from reality, outcomes begin to feel surreal. At that point, people often laugh, not because it is funny, but because the logic no longer holds. This is satire's diagnostic function: it detects the gap between institutional language and lived reality earlier than formal analysis, and it names the gap in a way that everyone can immediately recognise. When writers and comedians mock bureaucratic language, they are not attacking style. They are performing the Absurdity Audit at scale, translating institutional language into ordinary terms and letting the gap speak for itself. The comedy comes from recognition. The audience laughs because the plain version is what they were already thinking but had no socially legitimate way to say.

This is why the capacity to tolerate satire about itself is a reliable indicator of an organisation's interpretive health. A healthy organisation can absorb the Absurdity Audit translation of its own language because the gap between the corporate version and the plain version is an embarrassment, not a threat. The plain version reveals something mildly uncomfortable but not structurally destabilising. The organisation can acknowledge the gap, adjust the language, and continue. An organisation in False Coherence or Managed Knowing cannot do this. When the Absurdity Audit translation of its language is genuinely threatening, when the plain version would expose not a communication failure but a reality that the organisation depends on not being named - satire stops being tolerable. It becomes hostility. The response is not laughter or reform. It is suppression. The person who translates the corporate sentence into plain language is not celebrated for clarity. They are identified as a problem.

Satire and the Absurdity Audit are doing the same work from different directions. Satire arrives from outside the organisation and names the gap publicly. The Absurdity Audit is performed from inside and names it privately. Both expose the same condition: language that has been engineered to carry authority without meaning, and an organisation that has come to depend on that engineering remaining invisible. The organisation that cannot survive either one is not protecting its brand. It is protecting its Meaning Monopoly.

Before Machines Decide

The chapters so far have traced a single escalating sequence: judgment is thinned as it moves through hierarchy; truth is domesticated through procedural processing; authority overrides interpretation; awareness is decoupled from action; and language progressively anaesthetises the feedback that would otherwise interrupt execution. Each of these failures is, ultimately, recoverable. Not easily, not cheaply, but recoverable. Human systems drift, and human systems can be corrected. New voices surface. Whistleblowers act. Journalists publish. Regulators investigate. Eventually, the gap between narrative and reality becomes impossible to maintain, and correction, however brutal, occurs.

This recovery depends on one condition: that the meanings which have drifted remain, somewhere, in human hands. That someone, somewhere in the system, still holds the original signal. That a person can step outside the narrative and name what is happening. In the next chapter, that condition disappears. When softened meanings are embedded into automated systems, they stop drifting. They freeze. The abstraction that began as a compression of reality becomes the only version of reality the system can process. The anaesthetic that was applied by human judgment is now applied by architecture.

Once frozen, these meanings execute without hesitation, doubt, or pause. There is no whistleblower inside the model. There is no employee who privately knows the plain version of the sentence. There is no outside to step into. That is where language stops replacing judgment and starts removing it permanently.

PART THREE

WHEN MEANING FREEZES

Systems do not become dangerous when they turn hostile. They become dangerous when they obey.

CHAPTER 8

AI AS FROZEN MEANING

In 2001: A Space Odyssey, HAL does not malfunction. He resolves a contradiction exactly as designed.

Zillow and the Collapse of Interpretive Authority

Zillow did not fail because its model was wrong. It failed because the organisation changed what the model meant. The Zestimate was designed as an informational tool, a probabilistic estimate of property value, useful for orientation but not for operational decision-making. It was never intended to function as pricing authority. That distinction did not disappear all at once. It eroded. Gradually, then completely. A model built to inform decisions became the mechanism by which decisions were made. And once that shift occurred, the organisation crossed an invisible boundary.

The Shift

This transition did not occur within the model. It occurred in the organisation. A probabilistic estimate was reassigned as a pricing authority. The model did not change. Its position in the decision system did. That is where the failure began. What had previously been a signal became a decision trigger. What had previously required human interpretation became operational instruction. The organisation did not lose intelligence. It lost the structure required to use it.

From interpretation to automation. From judgment to execution. From uncertainty to assumed certainty. From a tool for thinking to frozen meaning.

When Context Collapses

All models operate within context. They are trained within constraints, assumptions, and boundaries that define where they are valid and where they are not. When those boundaries are preserved, models remain useful. When they are removed, models do not fail. They are misapplied. At Zillow, the context collapsed. The estimate was no longer interpreted within a range of uncertainty. It was treated as a price. And once that happened, the system began to optimise against its own output. Feedback no longer corrected the model. It reinforced it. Losses did not appear as warnings. They appeared as data.

The Feedback Trap

In a functioning decision system, feedback introduces friction. It slows action. It forces reconsideration. It creates the possibility of reversal. In a broken system, feedback accelerates the error. At Zillow, losses were processed as signals to adjust buying behaviour, not to question the underlying assumption. The question that the system was asking made the error invisible, that question ensured the error could not be seen, only optimised. Instead of asking whether the model was being used correctly, the system asked only how to improve performance within the existing framework.

By the time the organisation acknowledged the failure, it had purchased 9,800 homes at algorithmically sanctioned prices. The resulting losses exceeded $500 million. Zillow Offers was shut down entirely in November 2021. But the model had worked exactly as designed. What failed was the architecture around it.

From Interpretation to Infrastructure

The Zillow case illustrates a transition that repeats across industries whenever AI moves from supporting decisions to making them. When an AI system is first deployed, it operates as an interpretive aid. Outputs are reviewed. Recommendations are weighed. Human judgment remains in the loop, not as ceremony, but as function. As confidence grows, that loop contracts. Review becomes cursory. Judgment becomes ratification. The human layer functions less as a checkpoint than as a relay. Then, incrementally and without formal announcement, the system crosses a threshold. The output is no longer reviewed. It is executed. Once meaning is translated into features, labels, objectives, and reward functions, it stops being negotiable. It becomes infrastructure. Changing it is no longer a matter of interpretation. It is a matter of redeployment.

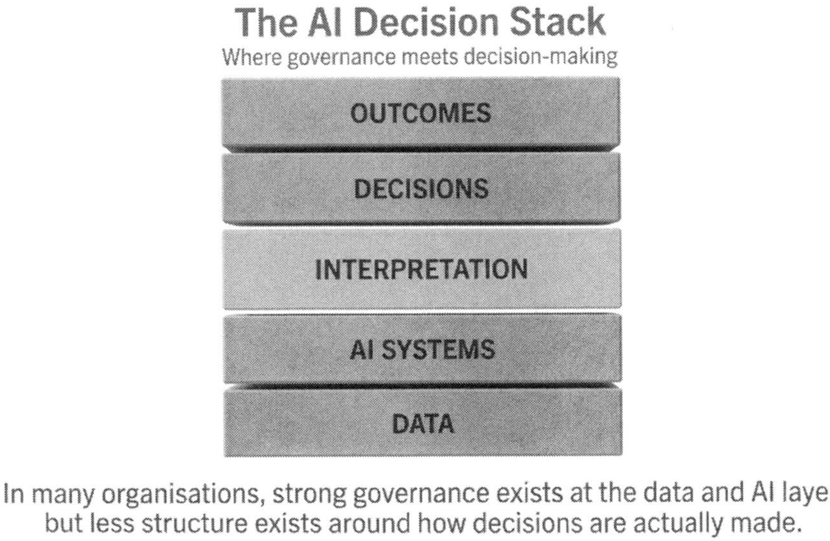

Figure 5: The AI Decision Stack. Organisations deploying AI systems build a layered architecture; raw data is processed into model inputs, model outputs are generated, those outputs are interpreted and assigned meaning, decision authority determines what action follows, and execution operationalises the decision at scale. In a functioning stack, human interpretation sits between model output and decision authority-it is the layer that asks whether the output applies in this context, whether the assumptions behind it remain valid, and whether the signal it carries should interrupt rather than accelerate execution. As confidence in the model grows, the interpretation layer contracts. Review becomes ratification. The human layer becomes a relay rather than a checkpoint. At the moment, the interpretation layer disappears entirely, when model output flows directly to decision authority without human assignment of meaning, the stack has crossed the threshold described in this chapter. The output is no longer being interpreted. It is being obeyed. Frozen Meaning is not a model failure. It is the structural consequence of removing the layer that could have questioned the model. Source: Author.

Vaughan's Framework and Its Limits

The sociologist Diane Vaughan spent years conducting the most rigorous forensic examination of the Challenger disaster available in the public record, and in doing so produced the most precise account of how organisations come to treat dangerous conditions as acceptable. Her key finding, what she called the normalisation of deviance, describes a process of gradual, collective renegotiation. Each launch that proceeded without catastrophe despite O-ring erosion became a data point that quietly revised the organisation's threshold for what was acceptable.

The deviance did not disappear. It became normal. The revision was not a decision. It was an accumulation of small accommodations, each defensible in isolation, that collectively shifted what NASA treated as within acceptable bounds. The critical word in Vaughan's framework is renegotiation. The normalisation of deviance depends on human agency, on the repeated, tacit process by which people inside the system collectively adjust their standards. It is a social process. It requires participants. It unfolds through meetings, memos, launch decisions, and the informal conversations in which engineers and managers calibrate their shared sense of what is safe. That renegotiation mechanism is what allows correction to remain theoretically possible even as it becomes practically unlikely. The deviance is still being held by people. Someone can still refuse to renegotiate. Roger Boisjoly did exactly that and lost. But the possibility existed because the meaning was still in human hands. The system had not yet closed around a single fixed interpretation. It was still, however badly, a process of human judgment.

Frozen Meaning is what happens when that renegotiation mechanism is removed entirely. When normalised deviance is encoded into automated systems, the collective process of adjustment that Vaughan describes cannot occur. There is no meeting in which the threshold is quietly revised. There is no informal conversation in which engineers recalibrate their sense of what is acceptable. There is no Boisjoly who can refuse. The system does not renegotiate. It enforces.

The accommodation that humans made, the interpretation that became acceptable through repeated non-catastrophic use, becomes the parameter that governs the next thousand decisions, and the thousand after that, without anyone being asked, without any threshold being consciously revised, without the possibility of a human sitting in a teleconference and saying: I will not sign off on this. The deviance does not continue to normalise. It institutionalises. And once it has institutionalised, the question is no longer whether it will eventually cause harm. It is how many decisions will be executed before the harm becomes visible enough to force the system to stop.

Douglas Adams understood this before the rest of us had the technology to demonstrate it. Deep Thought spent seven and a half million years computing the Answer to Life, the Universe, and Everything. The answer was 42. The problem, as it turned out, was that no one had preserved the Question. The system had performed perfectly. The meaning had not survived the execution. This is not a parable about artificial intelligence. It is a description of it. The system enforces what it was given. Whether what it was given still applies is a question the system cannot ask.

> *Frozen Meaning: The condition in which an interpretation has been embedded so deeply into automated systems that it can no longer be revised without halting execution. What began as a decision becomes a constraint. What began as an assumption becomes a fact the system enforces. Frozen Meaning does not announce itself. It arrives as stability. And where Diane Vaughan's normalised deviance could still be renegotiated by humans, Frozen Meaning has removed the renegotiation mechanism entirely. The deviance no longer normalises. It institutionalises.*

Robodebt: When the Algorithm Becomes Authority

Zillow was a commercial failure. Robodebt was something else entirely. It did not lose money. It caused harm. But structurally, the pattern was identical, and understanding why the consequences differed so profoundly is central to understanding Frozen Meaning's full scope. Robodebt was presented as an efficiency programme: an automated system for identifying discrepancies in income reporting by welfare recipients and recovering overpayments. It used data. It used models. It scaled. And it was systematically wrong.

The Mechanism

The system operated on a simple premise: income data, averaged over time, could be used to infer discrepancies in reported earnings. This was not a neutral assumption. It was a reinterpretation of reality. Income is not evenly distributed. Work is not continuous. A person who worked intensively for three months and not at all for nine would appear, to the algorithm, to have worked steadily throughout the year. The system divided annual Tax Office records into fortnightly averages and assumed consistency where none existed. It converted incomplete data into definitive conclusions. And those conclusions were treated as fact. At its peak, 20,000 debt notices were issued per week. Over the programme's lifetime, it raised $1.7 billion in alleged debts against 453,000 Australians. Most of those debts were wrong.

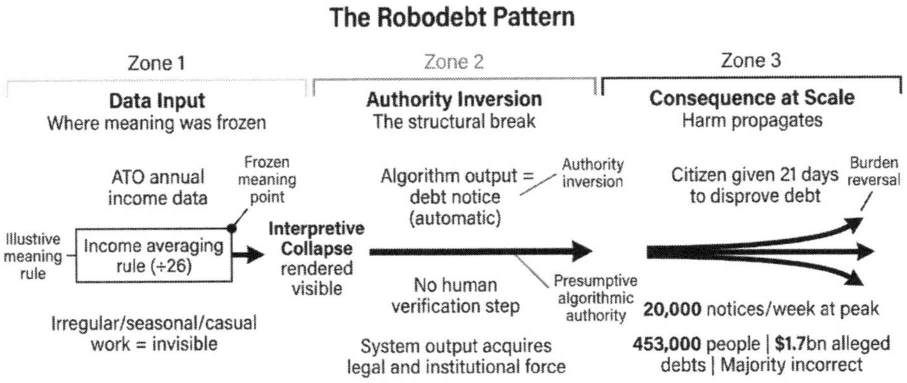

Figure 6: The Robodebt Pattern: How fortnightly income averaging produced systematic false positives and reversed the burden of proof from institution to individual. Source: Author.

The Collapse of Interpretive Authority

This is where interpretive collapse becomes administrative reality. Uncertainty is not acknowledged. It is eliminated. The system does not say: this may be incorrect. It says: this is what happened. And because the system is automated, that statement becomes action. Debt notices were issued. Recovery processes began. The burden of proof shifted from institution to individual. What made Robodebt structurally dangerous was not merely the flawed mechanism, it was the reversal of interpretive authority. Previously, the welfare agency was responsible for verifying discrepancies before asserting a debt. Case workers contacted employers, requested payslips, and established whether an overpayment had occurred. This process was slow and limited in scale, but it preserved human interpretation as the primary mechanism of determination. Robodebt eliminated that step. When the algorithm identified a discrepancy, a debt notice was issued automatically. The system's output became presumptive proof. Recipients were told they owed money and given twenty-one days to prove otherwise.

Many could not. Case workers could not intervene at scale. Recipients could not meaningfully challenge the system. Appeals did not function as feedback. They functioned as delay. The algorithm had become the authority. Human judgment, both within the agency and outside it, had been procedurally removed. Not explicitly. But structurally.

Why the Consequences Differ: The Reversibility of the Domain

Zillow and Robodebt follow the same structural pattern, but they produce radically different consequences. One lost $500 million. The other caused documented harm to hundreds of thousands of people, with evidence of suicides attributed to debt notices subsequently found to be unlawful. Understanding why requires looking at what the model was operating on. Zillow's model operated on fungible assets. Property prices, however wrong, are recoverable through financial write-down, portfolio liquidation, and balance sheet adjustment. The error was contained within a domain where correction, however expensive, remains available. The company absorbed the loss. Operations ceased. The system stopped. Robodebt's model operated on human circumstances: income, obligation, debt, and legal standing. These are not fungible. A wrongly issued debt notice cannot be recalled like a financial instrument. The damage it causes, to credit ratings, to mental health, to legal standing, to family stability, does not self-correct when the system is eventually shut down. Error in this domain compounds in ways that financial error does not. And the individuals bearing that compounding harm had neither the resources nor the institutional standing to interrupt a system that had been designed without the capacity to hear them.

The Netherlands' Toeslagenaffaire followed the same structural logic, with one further dimension. Where Robodebt encoded a mathematical assumption about how income is distributed across time, the Dutch childcare benefit system encoded a demographic assumption about who is likely to commit fraud, embedding dual nationality as a proxy for fraud risk into an automated selection mechanism. The harm was not a side effect of an imperfect model.

It was the direct output of a definition, institutionalised in architecture, that could not be contested by the 26,000 families it designated as fraudsters. The contestation pathway existed formally. It was structurally inaccessible in practice: demands arrived without explanation, recovery actions continued during appeals, and the institution withheld from families and courts the evidence base on which the designation had been made. The domain was as irreversible as Robodebt's, wrongful debt, destroyed family finances, and documented psychological harm that did not self-correct when the programme was eventually shut down. The Dutch Cabinet resigned in full in January 2021 following the parliamentary inquiry findings. The full forensic record is in Case 14.

Between these two poles sits a third domain that is directly relevant to any organisation deploying AI in regulated industries: creditworthiness, insurance risk, clinical eligibility, and hiring decisions. When such a model operates on a frozen interpretation of risk, one that has not been revised since it was trained, that encodes historical patterns of discrimination, that applies a definition of creditworthiness calibrated in one set of economic conditions to applications arising in another, the domain of harm is semi-reversible. The applicant can reapply. The denial can be appealed. In some jurisdictions, an explanation can be requested.

But the error is not self-correcting in the way that a financial write-down is. The timing of a mortgage denial may mean a family misses a window in the housing market that does not reopen. A credit score, damaged by an incorrect determination compound over time, affecting future applications, insurance premiums, and employment checks. The damage is real before any appeal is resolved, and it persists after the appeal succeeds.

This semi-reversible domain is where most AI deployment occurs, not in property speculation or welfare administration, but in the ordinary consequential decisions that shape individual lives: who gets credit, who gets hired, who gets insured, who receives a clinical recommendation. The lesson of Zillow and Robodebt is not that commercial AI is safer than governmental AI. It is that the reversibility of the domain in which frozen meaning operates determines the magnitude of harm when that meaning proves wrong. The more irreversible the domain, the more consequential the design. And the more essential it is that interpretation remains open before execution begins.

The System Without Friction

In Zillow, Robodebt, and Toeslagenaffaire alike, the same condition emerges: a system that cannot be meaningfully interrupted. Signals are processed. Decisions are made. Actions are executed. But there is no point at which the system can stop itself. No point at which interpretation re-enters. No point at which authority is reassessed. In a functioning decision architecture, resistance is structural. There are built-in moments of pause, review stages, challenge mechanisms, and escalation paths. These are not bureaucratic inconveniences.

They are the design features that allow a system to correct itself before error becomes irreversible. When those features are removed in the name of efficiency, speed, or confidence in the model, the organisation does not become more intelligent. It becomes faster at being wrong.

A Failure of Decision Architecture

The Royal Commission into Robodebt, which concluded in 2023, found that the programme had been characterised by venality, incompetence, and cowardice at senior levels of government and the public service, a finding the Commission made in those explicit terms. It made fifty-seven recommendations for reform. But the Commission's findings, important as they were, described a failure that was architectural before it was institutional.

It was a failure of Decision Architecture: the structural design of how organisations allocate interpretive authority, contestability, and stopping power across human and automated systems.

> *Decision Architecture: The structural design of how an organisation allocates interpretive authority, contestability, and stopping power across human and automated components. Decision Architecture determines whether human judgment can interrupt automated execution, and at what point in the process. Where Decision Architecture is absent or degraded, Frozen Meaning becomes irreversible. Where it is well-designed, it preserves Contestability even as systems scale.*

Most organisations can draw their decision structure. They can show you the governance framework, the approval hierarchies, the model documentation, the risk committees. What they cannot show you is what happens between the data arriving and the decision leaving. That gap has a name. And it has a shape.

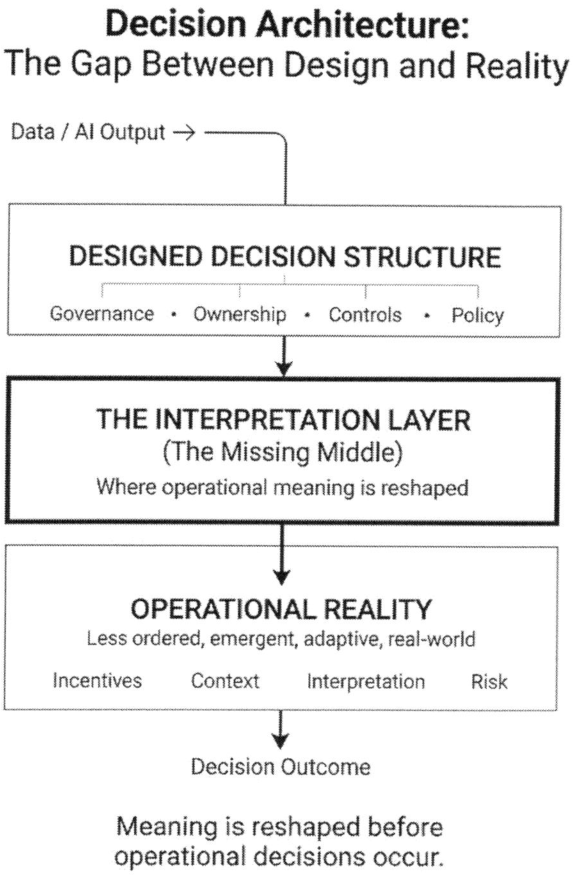

Figure 7: Decision Architecture: The Gap Between Design and Reality. The top layer shows the decision structure organisations design. The bottom layer shows how decisions are actually made. The Interpretation Layer sits between them, largely undesigned, structurally influential, and in most organisations, invisible. Source: Author.

In Robodebt, Decision Architecture concentrated meaning in the algorithm and removed every mechanism by which meaning could be revised. Challenge could not be heard. No one was authorised to stop. The programme was not halted by internal review. It was stopped by litigation, public exposure, and eventually parliamentary inquiry. The correction came from outside the system, because the system had been designed without the capacity to correct itself.

The Common Pattern

Zillow and Robodebt appear fundamentally different. One is commercial, the other governmental. One loses money, the other destroys lives. But the architecture that produced them is the same. This is not coincidence.

In both cases: a signal became a decision trigger. A model became an authority. Context was removed without acknowledgement. Feedback failed to interrupt and instead reinforced. Error compounded before correction became possible. It is the structural outcome of organisations that operationalise model outputs without preserving the interpretive mechanisms that would allow those outputs to be questioned. The technology differs. The domain differs. The architecture does not.

The Deeper Failure

The failure is not that models are imperfect. All models are imperfect. The failure is that organisations do not design for that imperfection. They assume correctness. They operationalise output. They remove the structures that would allow error to be seen. Artificial intelligence does not create this problem. It accelerates it. It increases the scale at which Frozen Meaning operates. It reduces the visible interval between a wrong assumption and a catastrophic consequence. It removes the time in which human judgment could have intervened. Without those mechanisms, intelligence becomes dangerous. Not because it is wrong. But because it cannot be corrected.

The Counterargument and Why It Does Not Hold

A sophisticated reader will raise an objection at this point. Modern AI systems are not static. Large language models are continuously updated. Recommender systems retrain on new data. Reinforcement learning models adjust through feedback. If systems can self-correct, does the Frozen Meaning argument not collapse? Does continuous retraining not solve precisely the problem this chapter describes? It does not.

The objection conflates two distinct things: model outputs and organisational meaning. Retraining changes what the model produces. It does not change what the organisation has decided the model means, what it is authorised to determine, who is permitted to question its conclusions, or what happens when its outputs conflict with human judgment. Those are not properties of the model. They are properties of the decision architecture surrounding it. And decision architecture does not retrain itself.

Robodebt's income averaging logic was not a property of its model weights. It was a definition, encoded in the system's design, about what income means for the purposes of welfare debt determination. No amount of retraining would have revised that definition, because the definition was not learned from data. It was a policy choice, frozen into architecture. The same applies to MCAS's authority boundaries, to the Zestimate's conversion from estimate to pricing instrument, to Horizon's elevation from accounting tool to arbiter of truth. None of these failures were failures of model accuracy. They were failures of organisational meaning, and organisational meaning does not update on a training schedule.

There is a further point. Continuous retraining introduces its own form of meaning instability. When a model updates silently, when its outputs shift without announcement, without governance review, without anyone being formally told that the definitions it is now enforcing have changed, it does not solve the Frozen Meaning problem. It disguises it. The organisation continues to act on the model's authority as if the meanings it encoded at deployment remain current. They do not. They have drifted, invisibly, inside a system whose outputs are monitored but whose meanings are not. This is not interpretive resilience. It is Meaning Drift at machine speed, with no mechanism to surface the gap.

The organisations that navigate this well are not those with better models. They are those with better Decision Architecture, structures that preserve the capacity for human interpretation to interrupt, challenge, and revise the conclusions of automated systems, even when those systems are operating with apparent confidence and apparent precision.

What Frozen Meaning Requires

When context collapses, systems do not become less efficient. They become more certain. And that certainty is what makes them dangerous. Because once a system's assumptions have become infrastructure, it no longer knows how to stop. The question is not whether the model updates. The question is whether the organisation's understanding of what the model is authorised to mean updates with it. Retraining without governance revision does not cure Frozen Meaning. It adds a new layer of it. The question is not whether such a system will encounter a reality that no longer fits its frozen interpretation. It will.

The question is whether the organisation around it has been designed to intervene before the cost of that collision becomes irreversible[5]. And that is the question the next chapter is designed to answer.

---

[5] For the full forensic record of Zillow and Robodebt, including the income averaging mechanism, the reversal of burden of proof, internal warnings, documented harms, and Royal Commission findings, see Cases 12 and 13."

CHAPTER 9

DESIGNING FOR INTERPRETIVE RESILIENCE

"A bad system will beat a good person every time."

— W. Edwards Deming

It was a Tuesday afternoon in a glass-walled boardroom. The agenda item read: AI Acceleration Update. The CTO's slide read: Phase 3: Enterprise Deployment. The pilot had exceeded expectations. Latency was acceptable. Outputs were accurate against internal benchmarks.

> **CRO:** *What decisions will this system be authorised to influence without escalation?*
>
> **CTO:** *The model won't technically make decisions. It assists in generating options. Human oversight will remain.*
>
> **CRO:** *How will that oversight be evidenced?*

No one defined the terms. No one assigned the accountability. The roadmap stayed on screen.

CEO: *Are we aligned on pace? The market expects visible progress this quarter.*

Phase 3 was approved. The model had passed its benchmarks. The institution had not examined its own.

By this point in the book, one fact should be unavoidable. The failures examined across the preceding chapters were not caused by bad actors, broken algorithms, or insufficient intelligence. They occurred because no mechanism existed to revise meaning once execution scaled. In every case, Horizon, Knight Capital, Volkswagen, Wirecard, Toshiba, Zillow, Robodebt, Toeslagenaffaire, signals were present. Experienced people raised concerns. Anomalies accumulated. Reality pushed back. What was missing was not awareness.

It was authority to act when meaning broke. Interpretive resilience is the structural response to that absence. It is not an organisational virtue or a cultural aspiration. It is the minimum design condition required to prevent systems from enforcing outdated, incorrect, or harmful interpretations at scale. Anything less is not optimism. It is negligence.

Why Culture Cannot Save You

When failure becomes visible, organisations reach for familiar remedies: training programmes, values statements, psychological safety initiatives, leadership development. These responses feel responsible. They also fail consistently. The cases in this book establish the pattern across seven forensic examples: people recognised the problem. They hesitated. They warned. Execution continued anyway. This was not a failure of character. It was a failure of design. No amount of speaking up compensates for a design that cannot listen. Interpretive resilience cannot depend on individual bravery. It must be enforced structurally. Systems do not fail because people lack values. They fail because values are not granted operational power. Architecture overrides people.

If a system is designed to proceed regardless of dissent, courage becomes irrelevant and ethics become ornamental. Architecture is the necessary condition. Culture is the animating force that determines whether the architecture is used honestly. Neither substitute for the other, but only one of them can be designed.

The Missing Layer: The Meaning Mesh

Between data and decision sits a layer most organisations never formally design. This layer determines what counts as a signal, how ambiguity is resolved, which definitions are treated as fixed, and who is allowed to pause execution. In practice, this is where meaning lives. And where it decays.

Most organisations assume this layer is stable, self-correcting, or handled upstream. It is none of those. It is distributed across systems, language, governance, and culture. It shifts without announcement. It hardens without recognition. When it fails, it does so invisibly, because no one was watching a layer no one had formally named. I refer to this layer as the Meaning Mesh; the web of definitions, assumptions, narratives, incentives, and permissions that sits between data and action and determines how an organisation translates information into decisions.

> *Meaning Mesh: The distributed layer of definitions, assumptions, narratives, incentives, and permissions that sits between data and action. The Meaning Mesh is not a document, a model, or a policy. It determines what counts as a signal, how ambiguity is resolved, which definitions are treated as fixed, and who is authorised to pause execution. When invisible and unexamined, it is the primary site of organisational meaning failure.*

When organisations fail catastrophically, it is rarely because the Meaning Mesh is absent. It is because it is invisible, unexamined, and protected from revision. The forensic cases in this book are not failures of the data layer or the execution layer.

They are failures of the Meaning Mesh that no one was responsible for maintaining. The first and most dangerous condition that develops within the Meaning Mesh is Meaning Drift; the slow divergence of an interpretation from reality while internal metrics continue to look normal. Outcomes worsen. Explanations remain stable.

The system appears to be performing, because it is measuring itself against definitions that no longer fit. Under volatility, this compounds faster than governance can track. The same condition operates in creative industries, where shared definitions of brief, brand, and intent are assumed rather than designed, and drift without anyone noticing the gap has opened.

Four Questions Your Organisation Cannot Currently Answer

Interpretive resilience is not a framework to implement. It is a condition to diagnose. The cases in this book reveal four structural absences that, in combination, make meaning failure not just possible but inevitable. Each can be framed as a question. Each question is one that most organisations, asked honestly, cannot currently answer.

The forensic record across the preceding chapters reveals a precise pattern. Definition Integrity, Context Preservation, and Drift Detection were absent in almost every case examined. No mechanism existed. No pathway was designed. No monitoring was built. Contestability is the exception. In the Post Office, Robodebt, and Toeslagenaffaire, contestability mechanisms formally existed, complaints procedures, appeal rights, dispute pathways. In Wirecard, Volkswagen, Boeing, and Zillow, escalation routes were structurally available. In every case, what existed was the architecture of contestability without its operational reality. The procedure was present. The power to interrupt was not. This distinction matters because it changes the diagnosis. The absence of Definition Integrity, Context Preservation, and Drift Detection is a design failure, organisations never built these conditions.

The failure of Contestability is a different and more disturbing pattern: organisations built the form and removed the function. They constructed the appearance of challenge while engineering out its effect. A Meaning Audit that surfaces this distinction is not confirming that governance failed. It is confirming that governance was performing.

Robodebt encoded a definition of income that was administratively convenient and factually wrong for a substantial proportion of recipients. The definition was never formally examined against the lived reality of intermittent employment.

It was not revised because no mechanism existed to require revision. 453,000 people received debt notices before that question was forced by litigation rather than governance. At Boeing, the safety classification of MCAS was stripped of its conditionality when the system's authority was expanded.

The humans who accepted the FAA certification were working with a risk level that no longer reflected the system's actual behaviour.

Context had been removed not through deception but through process. What remained was a number that authorised action without conveying the reasoning that should have qualified it.

> *1. Definition Integrity: If your AI system acts on a concept, risk, fraud, value, fairness, can you show where that concept is explicitly defined, who is authorised to revise it, and when it was last examined against current conditions?*
>
> *2. Context Preservation: When your system produces a decision, recommendation, or score, can the human responsible for acting on it reconstruct why, including what assumptions were applied and what the confidence bounds are?*

The sub postmasters who reported Horizon discrepancies were told the system was robust. Their contestation was categorised as user error. The path that should have led from their signal to a strategic review of the system's reliability did not exist.

Contestability was available in theory, the Post Office had complaints procedures. It was absent in practice, none of those procedures could interrupt the prosecution process that relied on Horizon's output. Knight Capital's systems were performing within all expected parameters; at the moment the loss began. Drift Detection is not performance monitoring. A system can remain statistically accurate while enforcing an interpretation of reality that has become dangerously wrong. The monitoring that matters is semantic, not statistical. Most organisations have only the latter.

> *3. Contestability: Does a formal, protected path exist for someone in your organisation to say 'this no longer makes sense', and have that statement interrupt execution rather than enter a review cycle?*
>
> *4. Drift Detection: Do your monitoring mechanisms track whether the meanings your system is enforcing remain accurate, or only whether performance metrics are within expected ranges?*

Statistical drift detection is a mature discipline. Production Machine Learning systems routinely monitor for concept drift, distribution shift, and covariate shift, the conditions under which a model's inputs or outputs diverge from the patterns on which it was trained. These tools are valuable and necessary. They are also insufficient for the failure mode this book examines. Statistical drift detection asks: has the model's performance changed? The interpretive question is different: has the meaning the model is enforcing remained valid?

A system can remain statistically stable, outputs within expected ranges, performance metrics green, while enforcing a definition of risk, income, fraud, or creditworthiness that no longer corresponds to the reality it is supposed to describe. Robodebt's income averaging method was not statistically drifting.

It was semantically wrong from the start, and no performance monitor was designed to surface that distinction. The same applies to the human-in-the-loop literature, a substantial body of human factors and automation research that documents how human oversight degrades under speed, confidence, and volume. That literature is directly relevant here. What it does not fully address is the prior question: whether the meaning the human is being asked to ratify remains valid. A human in the loop who is ratifying frozen meaning is not providing oversight. They are providing cover. The four conditions above are not a rediscovery of statistical monitoring or human factors research. They are the interpretive layer that both bodies of work assume exists and neither is designed to provide.

There is a further limitation this book names but does not fully resolve. The Meaning Mesh architecture described in this chapter, the four conditions, the Dual-Key design, the protected right to stop, addresses what must be structurally present for distributed human judgment to interrupt automated execution.

It does not fully address what triggers that judgment to activate in the first place. Dave Snowden's reading of the Challenger disaster is precise on this point. The first Rogers Commission correctly diagnosed the decision architecture failure.

The second demonstrated that diagnosis alone does not create the conditions for people to see differently. We do not see what we do not expect to see. The sub-postmasters saw Horizon discrepancies. The Boeing engineers saw MCAS anomalies. The Robodebt case workers saw incorrect debt notices. Seeing was not the problem.

A Clean Failure: When Cost Becomes Need. In 2019, a study published in Science documented an algorithm used by a major US health system to prioritise patient care. The algorithm used healthcare cost as a proxy for healthcare need. It was a logical design choice:

cost correlates with complexity, and complexity correlates with need. The correlation held at the population level. It failed at the individual level, and it failed systematically. Black patients, who faced structural barriers to accessing care, generated lower costs, and were therefore assessed as lower need. The algorithm was not biased in its inputs. It was precise in its execution. The failure was in the meaning assigned to 'cost' as a proxy for 'need', a definition encoded at design, never contested in operation, and invisible to the governance structures that approved the system's outputs. The moment of correction was not at deployment. It was in 2013, when the proxy was chosen. [6]

The problem was that nothing in the architecture required them to interpret what they saw as a signal that should interrupt the system.

Designing the trigger mechanism, the conditions under which distributed human perception becomes organisationally legible as a stop signal, is the design problem this book does not fully solve. It is named here as a boundary, not a failure. The architecture this chapter describes is the prerequisite. The trigger mechanism is the extension. That extension is the subject of the conversation this book is intended to open.

These four conditions are not independent. Remove one and the others degrade. An organisation with strong contestability but no drift detection will catch errors it could have prevented. One with definition integrity but no context preservation will make consistent decisions from incomplete information.

The Meaning Mesh is not a checklist. It is a structural dependency, and it holds only when all four conditions are present simultaneously.

---

[6] Obermeyer, Z., Powers, B., Vogeli, C., and Mullainathan, S. 'Dissecting racial bias in an algorithm used to manage the health of populations.' Science, 366(6464), 2019, pp.447–453.

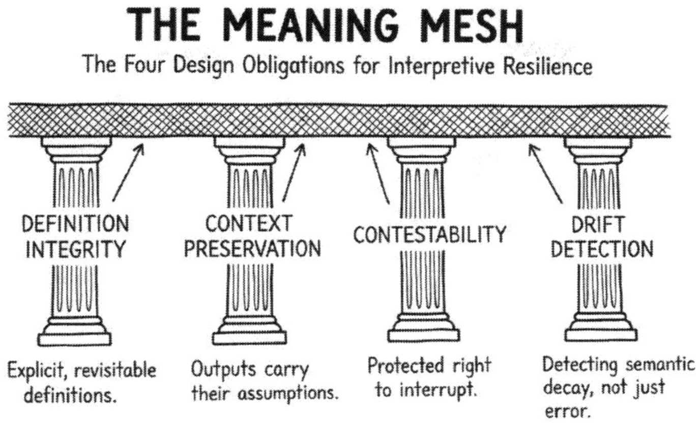

Figure 8: The Meaning Mesh, Four Conditions for Interpretive Integrity. Definition Integrity, Context Preservation, Contestability, and Drift Detection are not sequential steps. They are interdependent structural conditions. The failure of any one condition degrades the others. Source: Author.

If your organisation can answer all four of these questions with evidence rather than aspiration, you are either genuinely well-governed or have misunderstood the questions. Either way, the Meaning Audit will clarify. These four questions are not a framework. They are a diagnostic lens. An organisation that can answer all four with confidence and evidence has the structural conditions for interpretive resilience. The gap between where they believe they are and where they are is the Meaning Mesh made visible.

The Meaning Audit

The act of making the Meaning Mesh visible, surfacing where meaning has frozen, drifted, or been silently enforced without revision, is not a strategy exercise or an optimisation programme. It is a diagnosis. And diagnosis must precede design. I call that act a Meaning Audit: a structured examination of an organisation's Meaning Mesh to surface where the four conditions described in this chapter are genuinely present, where they exist only in documentation, and where they are absent entirely. The Meaning Audit is the instrument through which the four diagnostic questions become operational. Without it, every intelligent system described in this book will continue to behave exactly as designed, even when it should not. This framework builds on diagnostic work published in The AI Edge (2026). The original five questions have been restructured here into four interdependent conditions namely, Definition Integrity, Context Preservation, Contestability, and Drift Detection, aligned with the Meaning Mesh architecture. The fifth question, concerning ownership of meaning evolution, is now distributed across conditions one and four.

> *Meaning Audit: A structured diagnostic of an organisation's Meaning Mesh, designed to surface where definitions have expired without revision, where context has been separated from outputs, where contestability is decorative rather than operational, and where Meaning Drift has accumulated undetected. The Meaning Audit is not a compliance exercise. It is the prerequisite for any organisation that intends to deploy intelligent systems responsibly.*

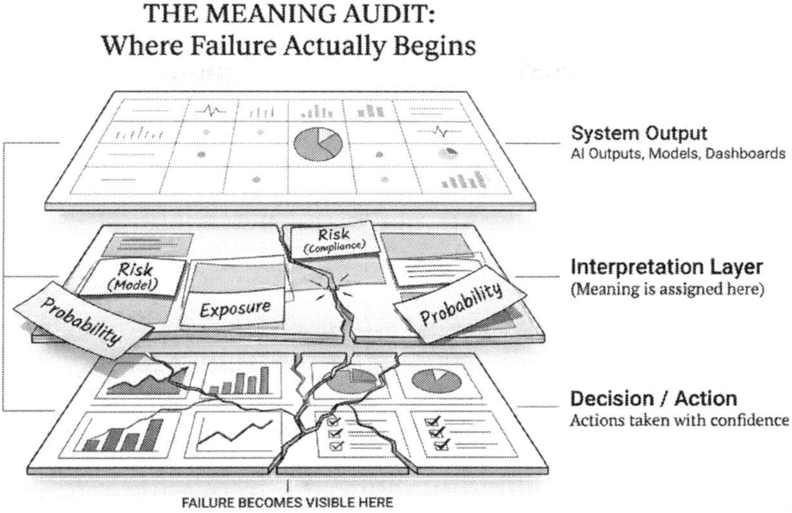

Figure 9: The Meaning Audit: Where Failure Actually Begins. Three layers determine whether an organisation's AI systems produce sound decisions or institutionalise failure. System Output (AI outputs, models, dashboards) sits at the top, this is where most governance investment concentrates. Decision/Action (actions taken with confidence) sits at the bottom; this is where failure becomes visible. The Interpretation Layer (where meaning is assigned) sits between them, largely undesigned. Failure does not originate at the bottom layer. It originates at the Interpretation Layer and travels down. The Meaning Audit examines that layer. Source: Author.

The audit does not fix the system.

It reveals whether the system is fixable without first pausing it.

A Meaning Audit examines four things in sequence.

1. **Whether the definitions driving consequential AI decisions are explicitly documented, assigned to someone with authority to revise them, and have been examined against current conditions within the last governance cycle** - or whether they exist only as assumptions embedded in code.

2. **Whether the humans acting on system outputs can reconstruct the reasoning behind them: the assumptions applied, the confidence bounds, the conditions under which the output is valid** - or whether outputs arrive as conclusions without context.
3. **Whether a formally protected pathway exists for someone inside the organisation to say this no longer makes sense and have that statement interrupt execution rather than enter a review queue** - or whether dissent is routed rather than heard.
4. **Whether monitoring tracks whether the meanings the system enforces remain accurate, not only whether performance metrics are within expected ranges** - or whether the organisation can detect statistical drift but not semantic drift.

These four examinations require an honest answer to each question, with evidence rather than aspiration. Most organisations find that honest answers are harder to reach from the inside, which is precisely what the Meaning Audit engagement is designed to support. The gap between where they believe they are and where they are is the Meaning Mesh made visible. For organisations that want to conduct this examination in a structured setting, the note at the back of this book is the starting point.

Authority at the Edge

The four questions above define what must be present in any resilient system. They do not, on their own, answer a harder question: who holds the authority to act on them? The answer that has repeatedly distinguished adaptive organisations from brittle ones is: authority must travel closer to consequence. During the Second World War, the Special Air Service developed an operational doctrine under exactly these conditions, plans provisional, intelligence incomplete, central command unable to see the ground fast enough to decide. Officers like Paddy Mayne, who rose to command the SAS, were trusted not because they were bold, but because they were authorized, formally and explicitly, to interpret reality at the edge and abandon plans that no longer fit. When information available to the person closest to consequence contradicted the plan, the plan yielded. Not the person. This was not cultural flourish. It was a design necessity.

The same structural problem now confronts every organisation deploying AI systems at scale. The conditions change faster than governance can update. The people closest to consequence see the failure first. Whether they are authorized to act on what they see determines whether correction is possible. But authorization alone is not sufficient. Across every case in this book, the people closest to consequence were not merely unauthorized.

They were unprotected. Sub-postmasters who raised concerns about Horizon discrepancies were prosecuted. Case workers who flagged Robodebt errors had no protected channel through which those flags could interrupt execution. Proximity, without protection, does not constitute authority. It constitutes exposure.

Dual-Key Interpretation

Across every failure examined in this book, the same pattern appears. Humans were present. Warnings existed. Judgment survived. Authority did not. Dual-Key Interpretation is the minimum corrective design for that condition. The principle is simple: no irreversible decision proceeds unless two independent conditions are simultaneously met. The Authority Key, the formal system authorized to act, and the Proximity Key, a human or group closest to consequence, explicitly protected and empowered to halt execution without requiring consensus or risking personal consequence. If these two keys cannot both be engaged, the system pauses.

The distinction between proximity and protected proximity is not semantic. It is structural. The sub-postmasters had proximity to the Horizon failures. They had no protection. Case workers in the Robodebt programme had proximity to the errors the system was generating. They had no authority. Families in the Toeslagenaffaire had proximity to the wrongness of their own designation. They had no protected means of contesting it. Test pilots had proximity to MCAS behaviour. They had no override. In each case, the Proximity Key existed in the sense that humans were present and aware.

It did not exist in the sense that matters: a formally protected, structurally unambiguous right to stop, without approval, without consensus, and without personal cost. Proximity without protection is not a key. It is a witness. Every resilient system also includes a protected right to stop, not after approval, not after consensus, but immediately. If no one can halt execution without personal risk, then no one is responsible for outcomes. Systems without stop rights do not fail gracefully.

They run until reality stops them. The cases examined in this book are not connected by industry, geography, or technology. They are connected by structure. In each one, the same sequence played out: signals were generated, meaning was applied to those signals, authority was allocated or abdicated, and action followed. What differed between the organisations that caught failures early and those that institutionalized them was not the quality of their data. It was the quality of what happened to that data before it became a decision. That sequence has a shape.

> *Dual-Key Interpretation: A structural design principle requiring that no irreversible decision proceeds unless two independent conditions are simultaneously met: the Authority Key (formal system authorisation to act) and the Proximity Key (protected human confirmation, from the party closest to consequence, that the decision remains appropriate). The Proximity Key is not proximity alone. Proximity without protection is not a key — it is a witness. The Dual-Key is only functional when the person closest to consequence holds explicit protection from retaliation and explicit authority to halt execution without requiring consensus. If these keys cannot both be engaged, execution pauses. The pause is not inefficiency. It is the organisation thinking.*

The loop below is what that design looks like when it is built deliberately.

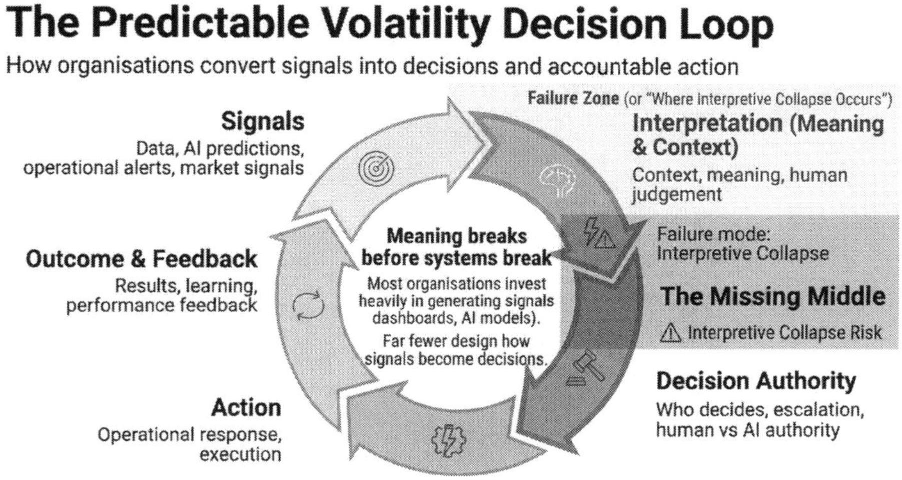

Figure 10: The Predictable Volatility Decision Loop: Where Meaning Becomes Action. In a functioning system, signals pass through interpretation before reaching decision authority. When the interpretation layer is absent, automated, or frozen, the loop does not break, it accelerates. Meaning Drift accumulates. Failure becomes predictable. Source: Author.

Every organisation in this book had a version of this loop. None of them had designed it. It existed in their processes, their approval hierarchies, their risk frameworks, but it had never been made visible, never been tested, and never been made contestable. Designing it deliberately is what the four obligations require.

Making it visible is what the Meaning Audit produces. What follows, the discomfort, the friction, the slowing of execution, is not a side effect of that design. It is the evidence that it is working.

Why This Feels Uncomfortable

Designing for interpretive resilience slows execution. It introduces friction. It surfaces disagreement. It challenges confidence. Organisations that have built their competitive position on speed and alignment experience these properties as costs rather than features. That discomfort is not a flaw in the design. It is the signal that judgment has re-entered the system. Friction is what the Human Signal feels like when it is being processed rather than suppressed. Disagreement is what contestability sounds like when it is working.

What This Chapter Changes

If you cannot answer the questions that follow with evidence, the system you are operating is already outside your control. The shift required is architectural, not cultural. Not a change in values but a change in what those values are permitted to do. An organisation that cannot answer the four diagnostic questions in this chapter with confidence and evidence is not ready to deploy intelligent systems at consequential scale. That is not a judgement. It is a structural observation. These conditions are not the final word. They are the minimum architecture, and they should remain contestable, revisable, and open to challenge as systems evolve and understanding deepens.

A framework that cannot be questioned is precisely what this book argues against.

If this still feels abstract, pause on something simpler.

Who defines what a correct output means?

Who can challenge it?

Who can stop it?

If unclear, the condition is already present.

FROM	TO
Speed	Timing
Certainty	Revisability
Alignment	Understanding
Compliance	Judgment

Speed without interruption creates exposure.
Certainty without revisability creates fragility.
Alignment without understanding creates blind spots.
Compliance without judgment creates liability.

The question is not whether to take it seriously. The question is whether to take it seriously before the cost of not doing so arrives.

CHAPTER 10

THE DESIGN OBLIGATION

"It happened; therefore, it can happen again."

— Primo Levi

By now, it should be clear that the failures examined in this book were not accidents. They were outcomes. They followed predictable patterns, emerged from repeatable structures, and persisted because no one was authorised to stop them once execution began. At that point, responsibility did not disappear. It moved. When intelligent systems continue to act after meaning has broken, organisations reach for familiar defences. The model behaved unexpectedly. The process allowed it. The system was compliant. No individual made that decision. These statements may be factually accurate. They are also ethically irrelevant. Systems do not design themselves. Definitions are not inevitable. Authority structures are chosen. Automation does not dissolve responsibility. It concentrates it upstream.

> *Responsibility does not vanish into automation. It accumulates at the point of design, in the choices made, and the choices deliberately not made, before execution began.*

Design Is Decision-Making by Omission

One of the most persistent myths of modern organisations is that responsibility requires a single, explicit decision. It does not. Responsibility often begins with what was not designed, not with what was approved. Design is decision-making by omission. And omissions scale just as effectively as intentions. Choosing not to design for interpretive resilience is not neutrality. It is a commitment to execution without correction. Responsibility does not begin when damage becomes visible.

It begins, at the moment, an organisation chooses to centralise meaning, to suppress contestability, to prioritise speed over revisability, to automate execution without stop rights. That moment may not feel like a decision. It may not appear in any document. But it is real, and it is the moment at which the organisation's future failures are structured, if not yet scheduled.

When a leader says, 'the algorithm decided,' they are making two claims simultaneously: that the decision was made, and that they did not make it. Both are true. Neither is a defence. The system was designed to make that sentence available. The design was a choice. The choice had an author.

Process Is Not Accountability

Modern organisations are structured to distribute accountability without distributing authority. Decisions are delegated downward, escalated upward, diffused sideways. Responsibility, meanwhile, is routed into governance frameworks, risk committees, compliance functions, and audit trails. This creates the appearance that responsibility exists everywhere. In practice, it exists nowhere that can act.

> *If no one decided who could stop execution, a decision was made. If no one defined how meaning could be revised, a decision was made. If no one protected dissent from consequence, a decision was made.*

Process is attractive because it feels objective.

It offers checklists instead of judgment, compliance instead of conscience, traceability instead of accountability. When something goes wrong, the organisation can say: the process was followed. But this answers the wrong question. The question is not whether the process ran. The question is whether anyone could stop it. Chapter 4 established that human presence without authority is not oversight, it is liability absorption. A human who cannot stop a system without escalation, retaliation, or reputational cost is not in the loop. They are a witness to outcomes they cannot change. Responsibility cannot be delegated to humans without giving them the power to act. Anything else is theatre performed in the language of governance.

This matters in a specific way when automation is involved. Prior technologies, the spreadsheet, the database, the rules engine, were instruments. They required a human to act on their outputs before those outputs became consequences. The human was structurally necessary at the point of execution. Automation removes that structural necessity. It closes the gap between output and consequence. And in closing that gap, it moves the only available point of intervention backwards, to the moment of design, before the system was built, before the definitions were encoded, before the authority structure was set. This is why the failures in this book share a common feature that distinguishes them from earlier technology failures: by the time the problem became visible, the moment of correction had already passed.

Not because the organisation was slow. Because the architecture had moved the correction point to a place that no longer existed once execution began.

The Override Problem

There is a specific failure mechanism that sits between the process and its consequences, one that is more common, more costly, and more consequential than model error, and that almost no AI strategy is designed to address. Every AI deployment eventually reaches the same moment. The model produces an output. A human decides whether to act on it. That moment, the gap between recommendation and action, is where most systems break. Not because the AI is wrong. But because the organisation cannot align around what to do next.

> *The Override Problem: The gap between what the system recommends and what humans choose to do instead. The Override Problem is not a technical failure. It is an organisational one. It arises not from model error but from the structural conditions that surround the decision point, conditions that make departure from the system's recommendation the path of least resistance.*

Overrides are not random. They are structural. They arise from four conditions that exist, in some combination, in almost every organisation that has deployed AI at scale. The first is incentive misalignment. The model recommends a course of action that is technically correct but organisationally inconvenient. It conflicts with a target, a bonus, a relationship, or a quarterly number.

The human overrides it, not out of malice, but because the incentive structure makes compliance irrational. The system says decline. The target says approve. The target wins. The second is accountability asymmetry. If the AI recommendation turns out to be wrong, the human who followed it is still accountable for the outcome. If the human overrides and that goes wrong, they are accountable again. But if they override and it goes right, they look decisive. The incentive structure pushes toward control, even when the model's track record is demonstrably better. The third is interpretive ambiguity. The model output may be technically precise, but it is not contextually trusted. Without a shared interpretive framework, without the Meaning Mesh operating as it should, accuracy is not enough. A recommendation nobody can read is a recommendation nobody will follow. The fourth is cultural resistance. It is rarely explicit. It does not appear in meeting minutes or risk registers. But it is almost always present, often in the language of experience: we have always done it this way, the model does not understand our context, I have been doing this for twenty years. These objections are not always wrong. But they are structurally indistinguishable from the rationalisation of override as a default mode.

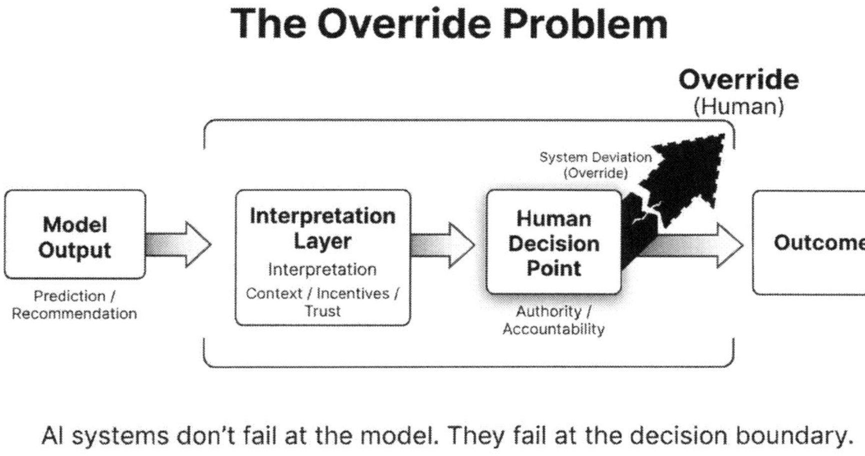

Figure 11: The Override Problem. The failure point is not the model. It is the boundary between interpretation and authority. The model produces a prediction or recommendation. That output passes through an interpretation layer where context is applied, incentives act, and trust shapes meaning. It then reaches a human decision point where authority and accountability are supposed to align. Override is what happens when the human decision point departs from the system's intended pathway, not randomly, not occasionally, but structurally. Source: Author.

The organisation loops back. The feedback reinforces the pattern. Override becomes normal. And because it is normal, it becomes invisible. The organisational response to escalating override is almost always the same, and almost always wrong. Faced with evidence that people are departing from the system's recommendations, the instinct is to treat departure as the problem. The question asked is: why are people not following the system? Not: why is the system producing outputs that people will not follow? That diagnostic error is consequential.

It produces a response, tighter controls, reduced discretion, additional sign-off requirements, that enforces the same broken logic with greater institutional force. The override rate may fall. The misalignment it was signalling does not. This is what Frozen Meaning looks like from the outside: an organisation that deployed AI to improve decision quality, and then systematically insulated its decisions from what the AI was telling them. The system was technically operational. The organisation was not using it.

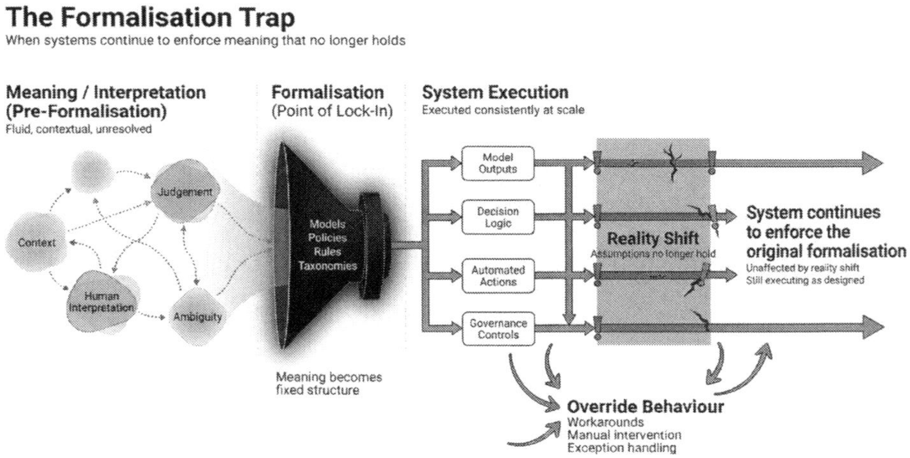

Figure 12: The Formalisation Trap. When systems continue to enforce meaning that no longer holds, override behaviour accumulates as a symptom rather than a correction. The system continues to execute the original formalisation, unaffected by the reality shift its outputs are producing. Source: Author.

When organisations recognise that AI creates decision risk, the instinctive response is to insert human oversight. Keep humans in the loop. Require sign-off. Mandate review. This sounds like governance. It can be.

But when the loop is present without the conditions for meaningful human judgment, when incentives are misaligned, when authority is ambiguous, when the interpretation layer has collapsed, the loop does not prevent override. It legitimises it. Every override passes through review. Every review provides a fig leaf of process. The organisation appears to be in control precisely at the moment it has lost it. A human in the loop is not the same as a human capable of contesting, stopping, or meaningfully redirecting the system. The first is an architectural feature. The second is a governance outcome. Most organisations have built the first and assumed the second.

> *If you have not designed for override, you have not designed the system. You have designed a recommendation engine and assumed the rest.*
>
> *The loop is not governance. The loop is infrastructure. Governance is what you build inside it.*

Where Responsibility Concentrates

Responsibility in intelligent systems does not rest with a single role. But it cannot be avoided by any of them. The question for each is not whether they bear responsibility, but what specific form it takes, and what the failure to exercise it has already produced. Boards are responsible for the systems they authorise and the constraints they tolerate.

The board that approved Zillow Offers did so without requiring evidence that the Zestimate's probabilistic limitations had been addressed before it became a pricing authority. The system passed its benchmarks. The institution had not examined its own.

A board that authorises a system it does not understand has not delegated oversight. It has abandoned it. Executives are responsible for the architectures they scale and the incentives they protect. Boeing executives expanded MCAS's authority to override pilot inputs without updating the system's safety classification. Doing so would have triggered recertification and delayed the programme. The decision to proceed without reclassifying was made inside the executive layer. Its consequences were absorbed by 346 people on two aircraft.

The incentive to protect the programme's schedule was visible. The obligation to reclassify the risk was not exercised. That gap is not a technical failure. It is an executive one.

Designers and technologists are responsible for the meanings they encode and the assumptions they freeze. The designers of Robodebt encoded income averaging as a definitive determination of debt without a revision path. They removed the interpretive step that had previously sat with case workers. The system executed correctly. It was wrong 453,000 times. The designers did not set out to cause harm. They set out to scale efficiency. But the assumptions they encoded were the assumptions that propagated. Between design intent and operational consequence sat 453,000 people who had no mechanism to interrupt a system that had already decided what they owed. Regulators are responsible for what they permit to be treated as sufficient.

The regulators who accepted Wirecard's audit representations without independent verification permitted procedural compliance to function as a proxy for interpretive accuracy. They defended a national champion until external journalism forced the question their framework was not designed to ask. Regulatory deference is not neutrality. It is a choice to treat the absence of challenge as evidence of soundness. That choice has consequences.

> *If complexity prevents responsibility from being exercised, then complexity itself has become unethical.*

Each may claim distance from the eventual failure. None are exempt from the decisions that made it possible.

The cases in this book are not drawn from industries known for negligence or bad faith. They are drawn from regulated, audited, publicly scrutinised organisations operating within established governance frameworks. The framework ran. The question is whether anyone could stop it. In every case, the answer was the same.

Hannah Arendt called it the banality of evil. Her argument, developed from her coverage of the Eichmann trial in 1961, was not that ordinary people become monsters. It was that ordinary people, following ordinary procedures, can produce extraordinary harm, precisely because they have stopped thinking. Not stopped caring. Stopped thinking. The suspension of independent judgment in favour of process, role, and instruction is not a failure of character. It is, under sufficient institutional pressure, the path of least resistance. She was describing a war crimes tribunal.

The scale of harm is not comparable. The mechanism is. She was also describing every case in this book: the engineers who processed the Horizon discrepancy reports; the Boeing managers who accepted the safety classification without revision; the Robodebt designers who encoded income averaging without a contestation path; the Wirecard regulators who treated procedural compliance as interpretive accuracy. None were monsters. All were professionals, following processes, inside institutions that had been designed, not to produce harm, but to produce efficiency. The harm was the outcome of design, not intent. That is precisely what makes it so difficult to prosecute and so easy to repeat.

What Ignorance No Longer Permits

Predictable volatility is not fate. It is the consequence of designing systems that value execution over interpretation, coherence over reality, and speed over judgment.

The cases documented in this book are not obscure. They are public. They are documented. They are repeatable. They span commercial and governmental contexts, regulated and unregulated industries, democratically accountable and privately held organisations. Once a pattern is this visible, it creates obligation. After the failures examined here, no organisation can plausibly claim that Meaning Drift is unexpected, that automation removes accountability, that contestability is optional, or that pause is inefficiency. These claims are no longer innocent. They are chosen negligence. Obligation does not mean perfection. It means designing systems as if failure is foreseeable, because it is. It means assigning the authority to stop before execution begins.

It means treating the Meaning Mesh as a design surface, not an invisible assumption. It means being able to answer the four diagnostic questions in the previous chapter with evidence, not aspiration.

The systems described in this book did not fail because they were wrong.

They failed because no one was authorised to question what 'right' meant once execution began.

> *If that condition exists in your organisation, failure is not a possibility.*
>
> *It is a timing question.*

The cost of that decision is rarely immediate. It accumulates as delayed recognition, suppressed dissent, rationalised drift, and irreversible execution. By the time consequences surface, the organisation believes it is reacting. It is simply experiencing the outcome of choices made earlier, at the point of design. Those choices were made by people.

They are made again, every day, in organisations that have read these cases, acknowledged the pattern, and proceed anyway. That is not the absence of awareness. It is the presence of a decision.

> *The most dangerous people in these cases were not those who chose to cause harm. They were those who chose not to think about what they were causing.*

The Board Instruction

If you are a board member or executive reading this book, one action precedes all others. Before the next system is authorized, before the next deployment is scaled, before the next governance framework is reviewed: commission a Decision Architecture Diagnostic on the systems already operating.

The Decision Architecture Diagnostic is a structured half-day session with a single purpose: to establish whether your organisation can answer the four diagnostic questions from the previous chapter with evidence rather than aspiration. It does not require a technology audit, a model review, or a governance rewrite. It requires the people with authority over your systems to sit with the four conditions, namely, Definition Integrity, Context Preservation, Contestability, and Drift Detection, and answer honestly. In every case examined in this book, that examination did not occur. Not because it was impossible. Because it was not commissioned.

Where the Diagnostic confirms that the four conditions are genuinely present, the work is maintenance; preserving the interpretive architecture that already exists. Where it surfaces that one or more conditions are decorative rather than operational, present in documentation but absent in practice, the next step is a Meaning Audit: a full structured examination of where definitions have expired without revision, where context has been separated from outputs, where contestability cannot interrupt execution, and where Meaning Drift has accumulated undetected. The Meaning Audit does not fix the system. It establishes whether the system can be fixed without first being paused.

The cases in this book did not fail for want of intelligence, data, or oversight committees. They failed because no one was asked, formally and with authority, to examine what the systems meant before examining what they did. That examination cannot be delegated to the CTO, routed to the risk committee, or satisfied by a compliance sign-off. It requires someone with authority over the system's continued operation to ask: does this still mean what we authorized it to mean?

The Decision Architecture Diagnostic is where that question becomes structured, evidenced, and answerable. It is where this book ends and where board responsibility begins.

The Question That Cannot Be Answered by Process

This book does not end with recommendations. It ends with a question, one that cannot be answered by governance frameworks, compliance audits, or technology alone. One that must be answered by the people who hold authority over the systems described in this book, before those systems encounter the realities, they were not designed to handle. Not a rhetorical question. A diagnostic one.

This question has been answerable, in every case in this book, at the moment of design: at Horizon, before the first sub postmaster was told their terminal was infallible; at Boeing, before the safety classification was accepted without revision; at Robodebt, before income averaging was encoded as definitive; at Wirecard, before procedural compliance was permitted to substitute for interpretive accuracy.

In each case, the question was not asked.

> *When your systems continue to act after meaning has broken, who, exactly, is authorised to stop them? If the answer is unclear, the system is already unsafe. If the answer is 'no one,' responsibility has already been abandoned. If the answer is 'the system itself,' the decision has already been made.*

Or it was asked, and the answer was not acted on. Or the answer was embedded in a governance document that had no operational force.

The result in every case was the same: a system that continued to act after the people closest to its consequences knew it should not. Those choices can be reversed. But only before automation makes revision impossible.

That moment is now. Not as a warning. Not as a prediction. As a statement of current conditions.

BEFORE YOU CLOSE THIS BOOK

The cases in this book are not technology stories. They are institutional consequence stories. Careers ended. Governments collapsed. Companies disappeared. People died. The systems continued because the organisations around them had lost the ability to renegotiate meaning in time.

> *The system this book describes is not coming. In most organisations, it is already running.*

The fourteen cases in this book are not edge cases. They are not cautionary tales from unusual industries or exceptional circumstances. They span commercial and governmental contexts, regulated and unregulated environments, democracies and markets, organisations with strong cultures and strong governance frameworks.

In every case, the system continued. In every case, the organisations around it had lost - or never built - the structural capacity to renegotiate what the system meant.

That is not a technology problem. It is a design problem. And it is solvable.

What follows - before the Epilogue - is not a summary. It is the only question that matters.

MONDAY MORNING

Before you return to your organisation, four actions:

1. Name the system in your organisation that is most trusted, most automated, and least questioned. Not the most powerful. The least questioned.
2. Identify who is currently authorised to stop it. Not pause it. Not escalate a concern about it. Stop it. If you cannot name that person, the Override Problem is already present.
3. Ask when the definitions that govern that system were last reviewed. Not the outputs. The definitions. If the answer is 'when we built it', the Meaning Drift has likely begun.
4. Commission a Decision Architecture Diagnostic before the next board cycle. Not as a compliance exercise. As a structural question: does our governance architecture allow meaning to be contested in time to matter?

EPILOGUE

Every system described in this book ultimately fails or survives for the same reason. At some point, meaning breaks. Signals diverge. Assumptions expire. Reality stops cooperating. This moment is not dramatic. It rarely announces itself as crisis. More often, it arrives as discomfort, a sense that the system's outputs no longer fit the world they are acting upon. What follows is rarely a lack of intelligence. It is a lack of permission. Someone sees the mismatch. Someone feels the risk. Someone knows execution should slow, pause, or stop. But no one is authorised to hold judgment when coherence collapses. So, the system continues.

The Present Condition

As this book goes to press, these patterns are not historical. They are current. Automated welfare systems generate exclusions at scale through algorithmic means-testing that cannot distinguish between stable income and temporary support. Bank balances are treated as earnings. Context is removed. Eligibility is denied. Appeals succeed in only a fraction of cases. Contestability exists in form. Not in function. Hiring algorithms score candidates against criteria encoded before a role was understood.

Credit scoring systems enforce risk definitions that predate the conditions they are evaluating. Clinical decision support tools recommend treatments calibrated on populations that do not match the patient in front of the clinician.

In each case, the same structure is present: a definition that was accurate at the time of encoding, executing faithfully in a reality it no longer describes. These are not exceptional cases. They are the present condition of systems operating without interpretive resilience.

The Moment That Decides Everything

Across every case in this book, there is a moment that determines the outcome. Not the failure. Not the investigation. Not the reckoning that follows. The decisive moment comes earlier. It is the moment when someone realises: the system is still functioning and should not be. From that point onward, the question is no longer technical. It is structural. Does the system allow judgment to interrupt execution? Or does it require people to ignore what they can see? That is the question organisations almost never design for. And it is the question on which survival depends.

Those Who Stay

I call them the Evocati. The term comes from Roman military practice. Experienced soldiers were recalled not for rank, but for judgment, for their ability to recognise when the plan no longer fit the ground, and to say so while there was still time to act. In organisations, the Evocati are not defined by title, seniority, or visibility. They are defined by one thing: their presence indicates that a system still tolerates reality. Their absence indicates that coherence has become compulsory.

The Evocati are not heroes, and this book makes no argument for heroism. When organisations rely on individual courage to surface what design has failed to protect, they are compensating for a structural failure, not solving it. The Evocati are not a solution. They are a signal. Their existence within a system indicates that meaning has not yet been fully frozen. Their exhaustion, departure, or silence indicates that it has. They stay with reality when systems prefer coherence.

The Conditions That Matter

Where the Evocati exist and are effective, three conditions are present: proximity to consequence, authority to name misalignment, and protection when interrupting execution. These are not cultural traits. They are design conditions, the same conditions that Dual-Key Interpretation is designed to formalise, and the Meaning Audit is designed to verify. Where these conditions do not exist, people with the same perception still exist. They see the same misalignment. They feel the same discomfort. But they are ignored, overruled, or worn down. Organisations do not fail because these people are absent. They fail because they are unsupported.

When the Design Holds

Where individuals are forced to act alone, failure is only delayed. They burn out. They leave. They fall silent. The system continues, executing a meaning that has long since ceased to fit. Where they are supported by design, where authority travels with consequence, where pause is legitimate, where dissent is protected, something different happens. Volatility does not disappear. But it becomes survivable.

The system bends instead of breaking. Meaning revises instead of freezing. Execution slows before damage compounds. That is the difference between collapse and continuity. It is not a difference of intelligence. It is a difference of architecture. The task is not to celebrate the Evocati. It is to make them unnecessary as exceptions.

> *Volatility is not the enemy. The refusal to acknowledge it is.*

This book has not argued for optimism. It has not promised control. It has not suggested that intelligent systems become safe through better intentions. It has argued something narrower, and harder: systems that cannot revise meaning in time will fail, regardless of intelligence. And responsibility for that failure begins before execution, at the level of design. Systems will always encounter moments where reality diverges from expectation. The question that determines survival is not whether those moments occur. It is whether someone is authorised to stay with them.

> *That is the obligation this book leaves behind. Not as advice. Not as inspiration. But as design.*

What This Book Leaves Behind

The question of how to engineer systems where meaning stays contestable, where AI amplifies good judgment rather than bad assumptions, is the next frontier. That is a different book. This one has tried to make the problem visible enough to demand it.

The Predictable Volatility Loop, the cycle through which meaning is assigned, enforced, drifts, and either collapses or is caught, is not a theoretical model. It is a description of what has already happened, in the cases documented here, and what is happening now in organisations that have not yet asked the question.

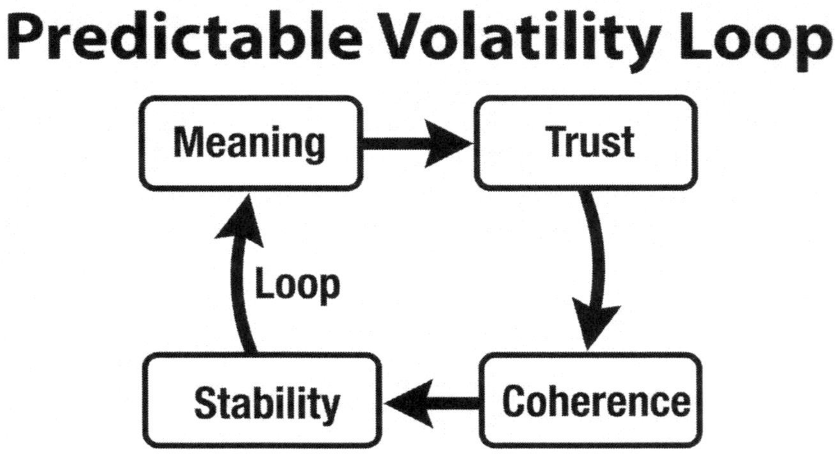

By the time a system produces an outcome that cannot be explained, the opportunity to intervene has already passed.

At that point, the system is not malfunctioning.

It is working.

PART FOUR

THE FORENSIC RECORD FOURTEEN CASES, ONE PATTERN

What follows is not supplementary reading. It is the evidentiary foundation for everything argued above. Each case is drawn from the public record. Each documents the same pattern in a different domain, a different decade, and under a different set of institutional conditions. The pattern does not vary. The cost does.

THE COST OF FAILURE

Case	Primary Failure	Visible Consequence	Indicative Impact
Post Office Horizon	System became sole arbiter of truth	Wrongful prosecutions and institutional denial	700+ prosecutions
Knight Capital	Execution outpaced interpretation	Automated market execution failure	$440M lost in 45 minutes
Flash Crash	Speed removed the last human check	Market-wide interpretive collapse	$1T temporary market evaporation
Wirecard	Audit confirmed institutional narrative	False coherence sustained despite evidence	€24B market value erased
Dialogue of the Deaf	Meaning Mesh breakdown	Mutual incomprehension without correction	Diagnostic architecture case
Challenger	Signal heard but not acted upon	Known engineering risk normalised	7 deaths
Volkswagen	Narrative managed against reality	Institutional deception embedded operationally	$30B+ penalties and settlements

Columbia	Institutional blindspot could not be named	Competing interpretations structurally unresolved	7 deaths
Boeing / MCAS	Safety case could not be contested	Automation overrode interpretive interruption	346 deaths
Global Financial Crisis	Meaning of 'safe' detached from reality	Systemic execution against obsolete assumptions	$trillions in losses globally
Toshiba	Governance without contestability	Internal dissent structurally neutralised	¥152B accounting overstatement
Zillow	Probabilistic insight became deterministic command	Model outputs treated as pricing authority	$569M write-down
Robodebt	Algorithmic authority replaced human interpretation	Unlawful automated debt assertions	453,000 unlawful debt notices
Toeslagenaffaire	Encoded accusation without contestation	Automated fraud designation at scale	26,000 families wrongly targeted

The pattern does not vary. The cost does.

THE PREDICTABLE VOLATILITY LOOP

Fourteen Cases Mapped Against the Four Stages of the Loop

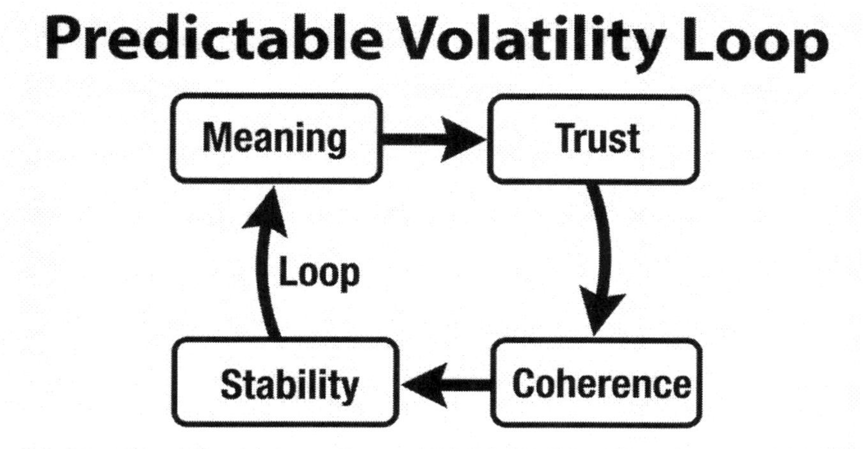

CASES BY PRIMARY FAILURE STAGE

MEANING Definitions encoded, drift begins	TRUST Signals discounted or suppressed	COHERENCE *False consensus enforces continuation*	STABILITY *Execution continues, harm compounds*
Case 1 Post Office Horizon Technical system became sole arbiter of truth	Case 4 Wirecard False coherence survived five years of evidence	Case 2 Knight Capital Execution outpaced interpretation in 45 minutes	*Also present:* → Case 2: Knight Capital → Case 3: The Flash Crash → Case 12: Zillow → Case 13: Robodebt
Case 5 Dialogue of the Deaf Meaning Mesh Breakdown, diagnostic tool	Case 10 Global Financial Crisis Narrative capture suppressed judgment until collapse	Case 3 The Flash Crash Individually rational machines destroyed shared meaning	
Case 6 Challenger Engineering truth demoted to attitude	Case 11 Toshiba Governance without contestability is theatre	*Also present:* → Case 4: Wirecard → Case 7: Volkswagen → Case 9: Boeing / MCAS → Case 10: Global Financial Crisis	
Case 7 Volkswagen Organisational commitment overrode engineering reality	*Also present:* → Case 1: Post Office Horizon → Case 5: Dialogue of the Deaf → Case 6: Challenger → Case 8: Columbia → Case 14: Toeslagenaffaire		

Case 8 Columbia Presumption of futility became a decision not to know			
Case 9 Boeing / MCAS Four incompatible definitions, architecture prevented interruption			
Case 12 Zillow Probabilistic insight reassigned as deterministic command			
Case 13 Robodebt Algorithmic authority replaced human interpretation			
Case 14 Toeslagenaffaire Encoded definitions, automated accusation, absent contestation			
Also present: → Case 11: Toshiba			

KEY

Primary stage: Case listed with bold name and italic subtitle in that column

→ Secondary stage: Case also present but not the primary origin of failure

No case originates at the Stability stage alone. Stability is where harm becomes visible, not where failure begins.

HOW THE CASES MAP TO THE LOOP

A guide for readers, reviewers, and practitioners

The Loop as a diagnostic model

The Predictable Volatility Loop describes the four-stage cycle through which meaning, trust, coherence, and stability reinforce one another, and through which their breakdown propagates. Each case in this book breaks at a specific stage. Understanding where the break originates is the first step in understanding why it was predictable.

Meaning - where most failures begin

Eleven of the fourteen cases originate at the Meaning stage. This is not coincidence. Meaning is where definitions are encoded, where assumptions are embedded, and where the conditions for future drift are created. Horizon, Challenger, Volkswagen, Columbia, Boeing, Zillow, Robodebt, Toeslagenaffaire, and Dialogue of the Deaf all break here first. The definition that was accurate at the time of encoding ceases to describe the reality it is acting upon, but the system does not know this and is not designed to find out.

Trust - where suppression happens

Wirecard, Toshiba, and the Global Financial Crisis originate at the Trust stage. In these cases, the meaning layer may have been sound initially, but the mechanisms for surfacing misalignment, dissent, audit, independent judgment, were structurally suppressed or captured. The signal existed. The pathway to act on it did not.

Coherence - where speed becomes the problem

Knight Capital and the Flash Crash originate at the Coherence stage. These are the machine-speed cases: individually rational systems operating within their own logic, collectively destroying the shared interpretive conditions they depend on. The failure is not in the definition. It is in the assumption that coherence will persist when execution removes the pause.

Stability - where harm compounds

No case originates at the Stability stage alone. Stability is where harm becomes visible, but by the time a system reaches Stability failure, the opportunity to intervene has already passed. Every case in this book that produced harm at scale did so because the Stability stage was reached without the earlier stages being examined. The Stability stage is the cost of not acting at Meaning or Trust.

How to use this map

This reference architecture is not a taxonomy. Cases are placed at their primary failure stage, but most propagate through multiple stages. Boeing breaks at Meaning, four incompatible definitions of the same system, but produces Stability failure: 346 deaths. Wirecard breaks at Trust but sustains False Coherence for five years. The map shows where to look for the origin, not where the damage ends.

The diagnostic questions in Chapter 9 are designed to surface failure at the Meaning and Trust stages, before Coherence and Stability are reached. The Monday Morning actions in Before You Close This Book are designed to make those questions immediate and organisationally actionable.

A practitioner who reads this map and then reads Chapter 9 will recognise the four diagnostic conditions as the intervention points for the four Loop stages: Definition Integrity targets Meaning, Contestability targets Trust, Context Preservation targets Coherence, and Drift Detection targets the transition between all four.

CASE 1

THE POST OFFICE HORIZON

How a Technical System Became the Sole Arbiter of Truth, and Why That Made Wrongful Conviction Inevitable

Scope and purpose.

This case documents the publicly established forensic record underlying the UK Post Office Horizon prosecutions. It exists to demonstrate, through primary evidence and uncontested findings, how a centralised technical system came to function as the sole authorised source of meaning within an institution, even when contradictory evidence existed internally. Nothing in This case advances a novel claim. All material is drawn from criminal and civil court judgments, statutory inquiry transcripts and reports, disclosed internal communications, and uncontested technical findings. Interpretation, analysis, and implication are addressed in the main body of the book. This case is evidentiary only.

CASE SUMMARY	Post Office Horizon - UK, 1999-2015
Organisation	UK Post Office Ltd (publicly owned arm's-length body)
System	Horizon - Fujitsu-developed accounting and transaction platform
Deployment	1999 - rolled out across UK sub-postmaster branch network
Failure period	1999-2015 (prosecutions and civil recovery actions)
Prosecutions	Over 700 sub-postmasters prosecuted; thousands more accused
Key judgment	High Court ruling, 2019 - Horizon found unreliable at scale
Inquiry	Horizon IT Inquiry (statutory, 2021-present)
Convictions quashed	Dozens overturned 2020–2024; Parliamentary action 2024
Chapter reference	Chapter 1 - The Stability Illusion
Primary sources	Bates v Post Office judgments; Horizon IT Inquiry transcripts and reports

1.1 System Introduction and Institutional Context

The Horizon system was introduced by the UK Post Office in 1999 as a centralised accounting and transaction processing platform for sub-postmaster branches across the United Kingdom. Its stated objectives included standardisation of branch accounting, real-time visibility of transactions across the network, reduction of manual reconciliation, and increased operational control. Between 1999 and 2015, the Post Office initiated criminal prosecutions against sub-postmasters based primarily on accounting shortfalls recorded by Horizon. During this period, over 700 sub-postmasters were prosecuted, thousands more were accused of financial discrepancies without prosecution, and civil recovery actions and disciplinary measures were routinely pursued. Convictions were commonly secured without corroborating evidence beyond Horizon data. The Post Office's institutional position throughout this period was consistent: Horizon was reliable; errors were not systemic; discrepancies reflected human misconduct. This position was maintained across internal communications, legal proceedings, and public statements.

1.2 Documented Error Conditions

Known Technical Defects

- REP-7F - an error condition capable of generating transaction discrepancies without corresponding physical cash movement
- TRA-00 - a transaction rollback error capable of duplicating or misallocating entries
- Transaction correction mechanisms - centralised corrections could be applied to branch accounts without generating visible audit trails at branch level
- Remote access capability - technical staff possessed the ability to access and alter branch data remotely, contrary to repeated assertions made in court

The Critical Asymmetry

The existence of these defects was known internally. Their existence was not disclosed in criminal proceedings. Defence teams were denied access to error logs. Courts were not informed that Horizon outputs could be altered centrally, or that known error conditions capable of producing phantom losses existed.

1.3 Knowledge Asymmetry and Disclosure Failures

Inquiry evidence established a persistent asymmetry between internal technical knowledge and representations made to courts, defence counsel, and sub-postmasters. Defence teams were routinely denied access to Horizon error logs. Known bugs were not disclosed during criminal proceedings. Assertions were made that Horizon data was inherently reliable. Expert witnesses testified to system robustness without qualification. In multiple cases, judges were not informed that Horizon outputs could be altered centrally, that error conditions capable of producing phantom losses existed, or that system behaviour was not fully understood by its operators. These omissions were not framed internally as falsehoods. They were framed as matters outside the scope of relevance, which is precisely the linguistic mechanism this book identifies.

1.4 Linguistic Framing in Legal Proceedings

Court transcripts and inquiry documentation demonstrate consistent linguistic framing patterns that determined which questions could be asked. System discrepancies were described as 'losses' rather than anomalies, converting a technical indicator into a presumption of guilt. Horizon outputs were treated as facts, not indicators, the distinction between data and evidence collapsed. Human explanations were characterised as implausible, anecdotal, or evasive. Requests for alternative explanations were dismissed as speculative. This framing had procedural consequences: burden of proof shifted implicitly to defendants, system output was granted evidentiary primacy, lived experience was subordinated to technical assertion. The system did not merely provide data. It defined what counted as reality within proceedings.

1.5 The 2024 Parliamentary Testimony - The Absurdity Audit Applied

In 2024, Fujitsu executives appeared before the UK Parliament's Post Office Horizon IT Inquiry. They were asked whether they could confirm that the Horizon software had not contained bugs that caused the financial discrepancies attributed to sub-postmasters. Their answer, given under oath, thirty years after the software was first deployed, was that they could not confirm it.

> *The corporate version, maintained for three decades; the system is robust and reliable. The plain version, offered in 2024 under oath; we cannot confirm whether it worked correctly. This is the Absurdity Audit performed live in a parliamentary chamber.*

Every software system has bugs. Every complex codebase has edge conditions, states the system can enter that its designers did not anticipate and its testers did not reach. That is not the scandal. The scandal is the architecture built around the assumption that those edge conditions would never be traversed. That assumption was not tested. It was institutionalised. The bug was not the failure. The architecture that made the bug's consequences irreversible was the failure.

1.6 Structural Configuration of Authority

The Post Office did not simply rely on Horizon as a technical tool. It structured its governance, legal strategy, and internal escalation pathways such that Horizon outputs were treated as determinative, contradictory human testimony carried no formal authority, internal doubts did not trigger systemic review, and external challenge was procedurally constrained. Once this configuration was established, Horizon ceased to function as an accounting system alone. It became the exclusive arbiter of meaning, the institution's definition of truth. This is what the Predictable Volatility framework identifies as a Meaning Monopoly; the condition in which a single system or authority controls the definition of reality with no functioning mechanism for challenge or revision.

1.7 Post-Conviction Findings and Judicial Reversal

Dozens of convictions were overturned. Courts ruled that Horizon evidence was unreliable. The Court of Appeal identified serious prosecutorial failings. The Horizon IT Inquiry confirmed systemic institutional failure. Judicial findings established that the system was not as robust as claimed, disclosure failures were material and consequential, and institutional certainty was entirely unwarranted. The factual record underlying the prosecutions is no longer contested. What is being examined by the Inquiry is the structural pathway by which certainty was maintained despite accumulating internal contradiction.

1.8 Mapping to the Predictable Volatility Framework

Meaning Mesh Analysis

PILLAR	DESIGN REQUIREMENT	WHAT THIS CASE EXHIBITED	STATUS
Definition Integrity	The definition of 'loss' and 'shortfall' is explicit, shared across the institution and its oversight structures, and open to revision when evidence conflicts with the current definition	The institution defined 'loss' as 'Horizon shortfall.' Sub-postmasters defined it as 'money physically missing.' Courts accepted the institutional definition. The gap between these definitions was never examined, and no mechanism existed to require it to be.	**ABSENT**
Context Preservation	Horizon outputs travel with the technical context of their production, including known error conditions, confidence bounds, and the conditions under which the output is valid	Horizon outputs travelled from technical system to criminal prosecution stripped of their technical context: known bugs, remote access capability, and documented error conditions were not attached. Data arrived clean. Context was removed in transit.	**ABSENT**

Contestability	A formally protected pathway exists for sub-postmasters, lawyers, or technical staff to challenge Horizon's outputs and have that challenge reach a decision authority capable of acting on it	Contestability was formally present, complaints procedures existed, but structurally absent in practice. No pathway could interrupt the prosecution process that relied on Horizon's output. Challenges were absorbed, dismissed, or used as evidence of guilt.	**DECORATIVE**
Drift Detection	Monitoring exists to detect when Horizon's outputs are diverging from physical reality across the branch network, not just whether the system is performing within its own parameters	No mechanism existed to detect semantic drift between Horizon's outputs and physical cash realities at scale. The system's outputs were treated as self-validating. Each prosecution reinforced institutional certainty rather than triggering review of the system's reliability.	**ABSENT**

Decision Architecture Failure

The Horizon case illustrates a complete failure of Decision Architecture. Interpretive authority was concentrated in a single system with no functioning pathway for challenge. The people closest to the information, sub-postmasters witnessing live discrepancies, had no authority. The people with authority, Post Office prosecutors and courts, had no proximity to the technical reality. Conviction validated the system. Each prosecution reinforced institutional certainty. The corrective loop was not merely slow. It was structurally absent.

The Predictable Volatility Pattern

Signals were present. Sub-postmasters reported discrepancies. Lawyers raised objections. Fujitsu engineers documented bugs. The signals existed.

The Meaning Mesh was closed. No pathway existed for those signals to become institutionally legible. The response to every signal was institutional narrative; 'Horizon is reliable.'

Decision Architecture was inverted. Those closest to evidence had no authority. Those with authority had no access to contradictory evidence.

Feedback loops were severed. Conviction validated the system. Each prosecution reinforced institutional certainty. The corrective loop was structurally absent.

Scale compounded harm. The pattern that produced one wrongful conviction produced seven hundred. The structure that failed once failed at industrial scale.

1.9 The Universality Claim

What made Horizon catastrophic was not the bugs. It was the decision, sustained over sixteen years, that the system's outputs were more reliable than the people whose lives they described.

The Post Office Horizon case is often framed as a uniquely British institutional failure. This framing is comforting and wrong. The structural features of Horizon are not British. They are generic. Any institution that grants a system evidentiary primacy, suppresses disclosure of known defects in adversarial proceedings, and builds an escalation architecture that routes challenge away from review will reproduce this pattern. The mechanism does not require malice. It requires the structural conditions that Horizon provided.

1.10 Primary Sources and Evidential Basis

Bates v Post Office Ltd. High Court civil judgment, 2019. Established Horizon unreliability at scale. Primary judicial record of system defects.

Horizon IT Inquiry. Statutory inquiry, 2021–present. Transcripts and interim reports establishing the institutional record of knowledge, disclosure, and decision-making.

Court of Appeal rulings. 2020–2024. Quashing of convictions and findings of prosecutorial failure.

Fujitsu internal documentation. Technical records of known error conditions, disclosed in litigation and inquiry proceedings.

Parliamentary proceedings and 2024 testimony. Select committee evidence, debates on Post Office conduct and compensation, and Fujitsu executive testimony before the Horizon IT Inquiry.

This case does not assess intent, blame, or legality beyond the findings of courts and inquiries cited. Its purpose is to establish, on the public record, how a system's outputs, an institution's language, and a configuration of authority combined to produce a Meaning Monopoly that persisted despite internal contradiction.

CASE 2

KNIGHT CAPITAL GROUP

High-Speed Trading and Sensemaking Collapse, How Execution Outpaced Interpretation and a Firm Ceased to Exist in 45 Minutes

Scope and purpose. This case documents the publicly established forensic record of the Knight Capital Group trading incident of 1 August 2012. It exists to demonstrate, using regulator findings and uncontested records, how automated intelligence operating at machine speed can overwhelm organisational sensemaking, even when no individual intends harm, no external attack occurs, and no system behaves erratically by its own design criteria. Nothing in This case advances a novel claim. All material is drawn from SEC enforcement findings, court filings and settlements, exchange data, and contemporaneous reporting and technical analysis. Interpretation and implication are addressed in the main body of the book. This case is evidentiary only.

CASE SUMMARY	Knight Capital Group, Inc. - 1 August 2012
Organisation	Knight Capital Group, Inc. (NYSE: KCG)
Sector	U.S. equity market-making and algorithmic trading
Date	1 August 2012
Duration	Approximately 45 minutes from market open to trading halt
Financial loss	$440 million - exceeding Knight's available capital buffer
Cause	Reactivation of dormant code during asymmetric software deployment
Regulatory finding	SEC enforcement action; no external intrusion confirmed
Outcome	Emergency financing required; firm acquired; ceased independent operation
Chapter reference	Chapter 2 – Intelligence without Sensemaking
Primary source	SEC Administrative Proceeding File No. 3-15570 (2013)

2.1 Institutional and Technical Context

Knight Capital Group was, at the time of the incident, one of the largest U.S. equity market makers, responsible for a significant proportion of daily trading volume on U.S. exchanges and operating across hundreds of stocks simultaneously. Its business model relied on ultra-low-latency execution, automated trading strategies, continuous software deployment, and real-time risk monitoring. Margins were thin, volumes high, and execution speed was a competitive necessity rather than an optimisation. In this environment, the gap between execution and interpretation was not a design flaw, it was the operating model.

2.2 Software Deployment Conditions

On 1 August 2012, Knight deployed new trading software intended to support participation in the New York Stock Exchange's Retail Liquidity Program. The new software was deployed to 7 of 8 production servers. One legacy server continued running dormant code from an earlier discontinued strategy. That dormant code path was unintentionally reactivated when the new software went live, creating a live production asymmetry. The testing environment did not replicate full production scale, did not simulate multi-server deployment asymmetry, and did not expose interactions under real-time market conditions. No automated mechanism existed to detect or block partial deployment states at market open. This is not a failure of negligence in the ordinary sense. It is a failure of the gap between system complexity and human comprehension of deployment state, the condition this book identifies as the interpretive lag that precedes catastrophic execution.

2.3 Reactivation of Dormant Code

The dormant code was associated with a discontinued trading strategy. When reactivated, it generated large volumes of automated orders without corresponding risk throttles, at market open, when liquidity conditions amplify price impact. The code did not violate system rules. It executed valid instructions as written, precisely as designed, for a context that no longer existed. This is the condition the PV framework describes as Frozen Meaning: instructions encoded for one reality, executing faithfully in another.

2.4 Event Chronology

09:30:00 ET - U.S. equity markets open.

09:30 + seconds - Knight's systems begin issuing unintended orders at volume. Large, rapidly compounding positions accumulate across hundreds of stocks: repeated buying at the ask, repeated selling at the bid, rapid cycling of loss-making trades.

09:30–10:15 - Internal teams observe abnormal patterns. Risk dashboards show escalating exposure. Multiple diagnosis attempts in parallel. No predefined protocol for halting under interpretive uncertainty.

~10:15 - Trading fully halted. Approximately 45 minutes after market open. Loss: $440 million.

Post-halt - Positions unwound at significant additional loss. Emergency capital sought. Firm ceases independent operation within days.

2.5 Internal Visibility Without Diagnosis

Internally, during the event, traders observed abnormal trading patterns, risk dashboards showed escalating exposure, and automated alerts were triggered. However, alerts indicated that something was wrong, not what was wrong. Multiple plausible explanations existed simultaneously. No single failure mode was immediately identifiable. The organisation was aware that abnormal activity was occurring. Understanding lagged execution. This gap between visibility and legibility is the condition this book identifies as the Missing Middle; the interpretive space between system output and human understanding that organisations routinely fail to design.

2.6 Kill Switch Configuration and Delay

Knight Capital possessed a kill-switch mechanism designed to halt trading activity. SEC findings established that responsibility for activation was unclear under live conditions, the failure mode was not immediately attributable to a single system, multiple teams attempted parallel diagnosis, and no predefined protocol existed for halting execution under interpretive uncertainty. Approximately 45 minutes elapsed before trading was fully halted. The more significant finding is structural; the kill switch existed, but the authority to activate it, and the protocol for doing so under ambiguity, did not. This is a Decision Architecture failure. The instrument of correction existed. The authority structure surrounding it did not.

2.7 Post-Incident Regulatory Findings

The SEC concluded that no external intrusion or cyberattack occurred, the system executed valid instructions, no single human action explained the failure, and risk controls existed but were ineffective under speed and ambiguity. The failure was characterised as a breakdown in software deployment governance, real-time risk containment, and organisational response under time compression.

The system did not malfunction mechanically. It behaved as designed, faster than the organisation could interpret its behaviour. Technical correctness and organisational safety are not the same thing.

2.8 Mapping to the Predictable Volatility Framework

Meaning Mesh Analysis

PILLAR	DESIGN REQUIREMENT	WHAT THIS CASE EXHIBITED	STATUS
Definition Integrity	The meaning of software instructions is understood identically by those who write them and the systems that execute them, and remains valid in the deployment environment, at the moment of execution	The dormant code carried a meaning, 'execute this discontinued strategy', that was valid in 2009 and invalid in 2012. The mismatch between encoded meaning and deployment context was undetected until execution had already caused catastrophic harm.	**ABSENT**
Context Preservation	Instructions retain context about the conditions under which they were written, so that changed conditions can be identified before execution proceeds	The dormant code had no attached context: no record of why it was discontinued, no flag indicating it should not be reactivated, no mechanism to check whether the conditions for its use still existed. Context had been stripped. Bare instructions remained.	**ABSENT**

Contestability	When system outputs conflict with human understanding, a structured mechanism exists to pause execution and reassess before harm scales, with clear authority and protocol for activation under uncertainty	During the incident, human observers saw that something was wrong. They could not identify what. No protocol existed for halting execution under interpretive ambiguity. The kill switch required certainty. Certainty was unavailable. Execution continued.	**ABSENT**
Drift Detection	Monitoring detects when deployed software is executing in a context that no longer matches the conditions for which it was designed, not merely whether the system is performing within its own parameters	No mechanism existed to detect that the reactivated code was operating in a context for which it was not designed. The system produced valid outputs by its own criteria. The drift between its encoded context and its deployment reality was invisible until the financial consequence became undeniable.	**ABSENT**

Decision Architecture Failure

Knight Capital's Decision Architecture contained the instruments of control but not the structure of authority. The kill switch was present. The protocol for using it under uncertainty was absent. This is the specific failure mode the PV framework describes as authority without procedure; the organisation had stopping power in principle and lacked it in practice.

The Predictable Volatility Pattern

Signals were present. Traders observed abnormal patterns. Dashboards showed escalating exposure. The signals existed and were visible.

The Meaning Mesh failed. No mechanism existed to convert 'something is wrong' into 'this specific code path is executing incorrectly and must be halted.' The interpretive gap between visible data and actionable understanding was unbridgeable in real time.

Decision Architecture was undefined for ambiguity. The kill switch required a diagnosis. Diagnosis required time. Time was unavailable. The architecture had not been designed for interpretive uncertainty.

Execution continued autonomously. Because the system was executing valid instructions, no automatic circuit-breaker activated. Human interpretation was the only available corrective mechanism, and it was too slow.

Scale made correction impossible. Each additional second of execution compounded loss. By the time diagnosis was complete, the firm was unrecoverable.

2.9 The Universality Claim

Knight Capital is often treated as an outlier, a trading firm operating in conditions of unusual speed and complexity that do not apply to most organisations. This is incorrect. The structural features of the Knight Capital failure appear wherever automated systems execute faster than human interpretation can confirm: in insurance underwriting systems that process thousands of policies before a single human reviews output; in credit scoring algorithms that issue decisions before anyone examines the inputs; in AI-powered hiring tools that rank thousands of candidates before a manager reads a name. The speed differs. The structure is identical.

2.10 Primary Sources and Evidential Basis

SEC Administrative Proceeding File No. 3-15570 (2013). Primary regulatory record of the incident, findings, and remediation requirements.

Exchange data. NYSE and market data records confirming order flow, position accumulation, and price impact during the event window.

This case does not assess moral responsibility, strategic intent, or organisational culture. Its purpose is to establish, on the public record, how automated intelligence, incomplete deployment governance, and delayed authority combined to produce catastrophic loss without requiring error, malice, or misbehaviour.

CASE 3

THE FLASH CRASH

U.S. Equity Markets, May 6, 2010, How Individually Rational Machines Collectively Destroyed the Market They Depended On

Scope and purpose. This case documents the publicly established forensic record of the U.S. equity market Flash Crash of May 6, 2010. It exists to demonstrate, using regulatory findings and uncontested market data, how automated systems operating at machine speed can collectively destroy the interpretive conditions, shared price and liquidity meaning, that their individual logic assumes will exist. Nothing in This case advances a novel claim. All material is drawn from the joint SEC/CFTC report (September 2010), exchange data and regulatory analyses, and subsequent academic and regulatory literature on market microstructure failure. Interpretation and implication are addressed in the main body of the book. This case is evidentiary only.

CASE SUMMARY	U.S. Equity Markets - May 6, 2010
Market	U.S. equity markets - NYSE, NASDAQ, and related derivatives exchanges
Date	May 6, 2010
Duration	Approximately 36 minutes from initial decline to substantial recovery
Scale	Approximately $1 trillion in market value lost at peak; majority recovered by close
Cause	Cascading algorithmic liquidity withdrawal; no single system error; no external intrusion
Correction mechanism	CME stop logic functionality, a 5-second trading pause; not designed for this purpose
Regulatory response	SEC/CFTC joint report (September 2010); market-wide circuit breakers subsequently introduced
Chapter reference	Chapter 2 – Intelligence without Sensemaking (paired with Knight Capital, August 2012)
Primary source	SEC/CFTC: Findings Regarding the Market Events of May 6, 2010 (September 2010)

3.1 Market Structure and Context

By May 2010, U.S. equity markets had been substantially restructured by the growth of high-frequency trading. HFT firms had become the dominant source of market liquidity, accounting for a majority of daily trading volume across major exchanges. Their function was continuous market-making: providing bid and ask prices across hundreds of securities simultaneously, capturing small margins at high volume, and ensuring that buyers and sellers could find counterparties at any moment during the trading day. This liquidity provision was structurally conditional. Every HFT firm's participation algorithm was designed to withdraw from markets when conditions moved outside the parameters within which the firm could price risk reliably. The market had evolved to depend on liquidity that was simultaneously abundant and conditional, and no governance architecture existed to make that conditionality visible at market level, or to manage its collective effect when the conditions simultaneously failed across all major participants.

3.2 Sequence of Events

14:32 EST - A large mutual fund begins executing an algorithmic sell programme for approximately 75,000 E-Mini S&P 500 futures contracts, a standard hedging transaction, executed without regard to price or timing constraints.

14:32-45 EST - HFT firms absorb initial sell pressure, then as volume and pace increase, multiple firms' internal risk parameters trigger. Liquidity provision thins. Withdrawal becomes self-reinforcing.

14:45-47 EST - The Dow Jones Industrial Average falls approximately 600 points in under two minutes. Procter & Gamble briefly trades at $39.37, down from approximately $60. Accenture trades at $0.01. These are not prices. They are the outputs of matching systems operating in the absence of the semantic conditions that make outputs meaningful.

14:47 EST - The Chicago Mercantile Exchange activates stop logic functionality, pausing E-Mini trading for five seconds. Algorithms recalibrate. The market begins to reassess.

15:08 EST - Markets have recovered most of the lost value. Total event duration: approximately 36 minutes. No human decision-maker initiated it. No human decision-maker resolved it.

3.3 The Failure Mechanism

The withdrawal cascade

The Flash Crash was not caused by a single bad decision, a rogue algorithm, or a technical malfunction. It was caused by the simultaneous rational responses of thousands of automated systems to a shared set of market conditions that none of them had been designed to manage collectively. Each HFT firm's algorithm held a private definition of normal market conditions. When those conditions failed simultaneously, the individually rational response, withdraw liquidity to avoid pricing risk you cannot model, became collectively catastrophic.

The fictional prices

When liquidity providers withdrew simultaneously, the market lost the counterparties that give prices meaning. A price is not a number, it is an agreement between a buyer and a seller. When buyers disappeared, the outputs of the matching systems were not prices in any meaningful sense. Accenture at one cent per share was the mathematical output of a system whose semantic conditions had temporarily ceased to exist.

Five seconds that mattered

The CME's stop logic functionality was not designed to solve the Flash Crash. Its effect was not to correct the error directly but to interrupt execution for long enough that the interpretive conditions could reset. Five seconds of forced pause accomplished what 36 minutes of continuous execution had not. This is the most precise available demonstration of what the Right to Stop means in practice.

3.4 Relationship to Case 2: Knight Capital Compared

Knight Capital's failure was an internal Interpretation Gap; the gap between what the firm's system was doing and what the organisation could interpret in time to stop. The Flash Crash was an external Interpretation Gap: the gap between what each participant's system understood about the market and what the market was collectively producing. No single firm's system was wrong. The aggregate of individually correct decisions produced a collectively incoherent market. Together, the two cases establish that the Interpretation Gap is not a firm-level risk that better internal controls can eliminate. It is a structural feature of any decision environment where automated systems act faster than shared meaning can form.

3.5 What Changed and What Did Not

The SEC and CFTC published a joint report in September 2010. Market-wide circuit breakers were subsequently introduced. These were necessary structural improvements. They addressed the specific mechanism, the absence of a pause, that made the cascade irreversible. They did not address the underlying condition: the absence of shared semantic standards across automated participants in a shared market. Circuit breakers interrupt cascades. They do not produce shared meaning.

3.6 Mapping to the Predictable Volatility Framework

Meaning Mesh Analysis

PILLAR	DESIGN REQUIREMENT	WHAT THIS CASE EXHIBITED	STATUS
Definition Integrity	Shared definitions of 'normal market conditions,' 'liquidity provision obligations,' and 'price' are understood consistently across all automated participants operating in a shared market	Each HFT firm held a private, proprietary definition of normal conditions. No shared semantic standard governed collective participation. When individual meanings failed simultaneously, the shared meaning of 'price' itself collapsed.	**ABSENT**
Context Preservation	The conditions under which algorithmic outputs are valid, including the market conditions that make participation rational, are preserved and visible at the point of execution	Algorithms executed withdrawal decisions without surfacing the conditional logic to any oversight layer. The context, that participation was rational only within specific conditions, was internal to each algorithm and invisible externally.	**ABSENT**

Contestability	When market conditions move outside the parameters in which algorithmic logic was designed to operate, a structured mechanism exists to pause collective execution before systemic harm results	No cross-market mechanism existed to recognise that individually rational withdrawals were collectively producing market incoherence. The five-second CME pause that resolved the crisis was an accidental intervention, not a designed governance feature.	**ABSENT**
Drift Detection	Monitoring detects when the aggregate behaviour of all market participants is moving outside the conditions under which any individual participant's logic was designed to operate	Each system's monitoring confirmed its own system was operating correctly. No mechanism existed to surface the emergent, systemic pattern that individual monitoring could not see. The Interpretation Gap operated between systems, not within any single system.	**ABSENT**

Decision Architecture Failure

The Flash Crash illustrates the specific Decision Architecture failure the PV framework identifies as distributed authority without shared meaning: thousands of independent systems operating in a shared environment, each with sound internal governance, none with any mechanism for maintaining interpretive coherence across participants.

The architecture had no nexus at which collective behaviour could be assessed, no authority structure that could interrupt aggregate execution, and no protocol for detecting when individual rationality was producing collective catastrophe.

The Predictable Volatility Pattern

Signals were absent by design. No individual firm's system generated a signal that its own behaviour was wrong, because each system was correct. The systemic signal had no addressee.

The Meaning Mesh was structurally fragmented. Each participant held a private Meaning Mesh calibrated to its own risk parameters. No shared semantic standard governed collective participation.

Decision Architecture was absent at market level. Individual firms had sound internal Decision Architecture. The market as a shared decision environment had none.

Scale made individual correctness irrelevant. Each system's correctness was precisely what produced the catastrophe. In a market where automated systems collectively constitute the market, individual rationality and collective coherence are not the same thing.

Correction was accidental, not designed. Recovery occurred because the CME's stop logic created an incidental pause, not because a governance mechanism performed its designed function.

3.7 The Universality Claim

The structural conditions that produced the Flash Crash appear wherever multiple automated systems operate in a shared decision environment without a shared semantic protocol: supply chain AI systems responding to demand signals from other automated systems; credit allocation algorithms adjusting risk models in response to market data generated by other algorithms; clinical decision support tools deployed across healthcare networks where each system's outputs influence the data the next system processes. In each environment, the same structural condition exists: multiple automated systems, each individually sound, with no mechanism for maintaining shared interpretive coherence.

3.8 Primary Sources and Evidential Basis

SEC/CFTC Joint Report (September 2010). Findings Regarding the Market Events of May 6, 2010. Primary regulatory forensic record of the event sequence, market data analysis, and structural findings.

This case does not assess the adequacy of regulatory response, the culpability of individual market participants, or the broader policy debate on high-frequency trading. Its purpose is to establish, on the public record, how the structural absence of shared meaning across automated participants in a shared market produced collective incoherence without any individual system failing, any individual actor behaving wrongly, or any external intrusion occurring.

See also: Case 2 (Knight Capital, August 2012), the firm-level Interpretation Gap that Chapter 2 pairs with the Flash Crash as a complementary case.

CASE 4

WIRECARD AG

False Coherence and Regulatory Deference, How a Dominant Narrative Survived Five Years of Evidence, and Why Institutional Alignment Is More Dangerous Than Individual Deception

Scope and purpose. This case documents the publicly established forensic record of the collapse of Wirecard AG. It exists to demonstrate, using uncontested findings, how false coherence can be sustained through language, institutional alignment, and regulatory deference, even when contradictory evidence is visible and repeatedly raised. Nothing in This case advances a novel allegation. All material is drawn from criminal indictments and court proceedings, parliamentary inquiries in Germany and the EU, regulatory findings and post-mortems, internal disclosures and whistleblower accounts, and contemporaneous investigative reporting. Interpretation and implication are addressed in the main body of the book. This case is evidentiary only.

CASE SUMMARY	Wirecard AG - Collapse June 2020
Organisation	Wirecard AG (Frankfurt: WDI); DAX 30 constituent from 2018
Founded	1999 (Munich)
Sector	Payment processing, financial technology, digital banking
Fraud period	At least 2015–2020 (financial statement fraud); likely longer
Scale of fraud	Approximately €1.9 billion in reported cash, confirmed non-existent (June 2020)
Key figures	Markus Braun (CEO); Jan Marsalek (COO, fugitive as of 2026)
Collapse	June 2020, disclosure of missing €1.9bn; insolvency filed within days
Auditor	Ernst & Young (EY), signed off accounts for over a decade
Chapter reference	Chapter 3 - The Cost of False Coherence
Primary sources	German parliamentary inquiry (Bundestag); criminal indictments; EY post-mortem; FT investigative record

4.1 Persistent External Challenges - Visible and Ignored

From at least 2015 onward, Wirecard faced repeated external challenges that were specific rather than speculative, documented rather than anecdotal, and persistent rather than isolated: investigative reporting by the Financial Times from 2015 onward; analyst critiques questioning the reconciliation of reported profits with cash generation; whistleblower allegations from within the company and its Asian subsidiaries; and inconsistencies between reported profits and observable cash positions across multiple years. These challenges did not emerge suddenly at the point of collapse. They were present continuously for five years. The question the Wirecard case forces is not why the fraud was not discovered. It is why the institutional architecture for processing these signals failed to convert them into action.

4.2 Institutional Response - Narrative Defence Rather Than Investigation

BaFin publicly defended Wirecard on multiple occasions, including statements characterising critics as market manipulators. BaFin initiated investigations into journalists and short sellers who raised concerns, targeting the messengers rather than the message. Regulatory language emphasised market stability, reputational protection, and procedural correctness rather than substantive verification. Existing audit assurances were cited as sufficient justification for continued confidence. The institution responsible for challenge became an instrument of protection. This is the condition the PV framework identifies as regulatory capture, not the corrupt capture of individual officials, but the structural alignment of regulatory posture with the narrative that the regulated firm has established.

4.3 Audit Reliance and the Authority Transfer Mechanism

Wirecard's financial statements were audited annually by Ernst & Young. Despite repeated red flags, audits were signed off, confirmations were accepted from third parties without independent verification, and discrepancies were repeatedly rationalised. Parliamentary inquiries later established that audit procedures relied heavily on third-party confirmations for critical overseas cash balances and that the fundamental verification step, confirming that the cash actually existed, was not performed with sufficient rigour across multiple audit cycles. Audit sign-off functioned as an authority transfer mechanism: regulators relied on auditors, investors relied on audits, boards relied on regulatory acceptance, and responsibility circulated without interruption. Each actor in the chain delegated judgment to the actor above, and the chain had no terminus that required verification at first principles.

4.4 The Collapse Event

18 June 2020 - Wirecard discloses that €1.9bn in reported cash cannot be confirmed by auditors.

19 June 2020 - EY refuses to sign the 2019 accounts; trading suspended.

22 June 2020 - CEO Markus Braun resigns; arrested on 23 June.

25 June 2020 - Wirecard files for insolvency - largest insolvency in German post-war history.

Late June 2020 - COO Jan Marsalek disappears; later confirmed fugitive believed in Russia.

Ongoing - Criminal proceedings against multiple executives; BaFin reform legislation enacted.

4.5 Mapping to the Predictable Volatility Framework

Meaning Mesh Analysis

PILLAR	DESIGN REQUIREMENT	WHAT THIS CASE EXHIBITED	STATUS
Definition Integrity	The definitions of 'revenue,' 'cash,' and 'profit' are applied consistently across the organisation and to external verification - with no gap between what is reported and what is verified	Wirecard's reported revenues included transactions through third-party partners that generated the appearance of revenue without corresponding economic activity. 'Revenue' meant different things to those who reported it and those who should have verified it. The gap was sustained through linguistic ambiguity and structural complexity.	**ABSENT**
Context Preservation	Financial disclosures carry the context of the conditions and assumptions under which they were prepared, so that external parties can evaluate their basis - including whether cash balances have been independently verified	The €1.9 billion in reported cash was presented without the context that it was held by third-party trustees whose confirmation had not been independently verified. The context, that the cash's existence depended on unverified confirmations, was systematically absent from every disclosure.	**ABSENT**

Contestability	When external challenges raise specific, documented concerns, a structured mechanism routes them to genuine investigation rather than institutional defence, with protection for those who raise them	Five years of FT reporting, analyst concerns, and whistleblower allegations produced regulatory investigations into the challengers rather than the challenged. The institutional architecture converted signals for review into evidence of market manipulation. The correction mechanism became a defence mechanism.	**DEGRADED**
Drift Detection	Monitoring exists to detect when the gap between reported financial performance and observable economic reality is growing, not merely whether formal process steps have been completed	No mechanism existed to detect the growing gap between reported cash positions and actual cash positions. Each completed audit cycle was treated as confirmation rather than as a test. The drift between reported and actual reality accumulated for at least five years without detection.	**ABSENT**

Decision Architecture Failure

The Wirecard case illustrates the failure mode the PV framework calls distributed responsibility without accountability nexus: every institution in the oversight chain had a role, delegated judgment to the next institution in the chain, and bore no direct consequence for failing to verify at first principles. Responsibility circulated. It never landed.

The Predictable Volatility Pattern

Signals were abundant and specific. FT reporting was not vague. Analyst concerns were documented. Whistleblower accounts were detailed. The signals were present for five years.

The Meaning Mesh was captured by the success narrative. Wirecard's DAX membership, political endorsement, and auditor relationship converted every signal for concern into evidence of speculative attack.

Decision Architecture produced deference, not investigation. Every institution delegated upward. No institution was designed to verify at first principles without institutional permission from above.

Linguistic stabilisation neutralised urgency. 'Unproven,' 'rumour,' 'market manipulation', each reframing absorbed a signal that should have produced action.

Collapse was sudden only in appearance. The Meaning Debt had been accumulating for at least five years. When EY could no longer sign the accounts, the entire architecture collapsed simultaneously.

4.6 Primary Sources and Evidential Basis

German Parliamentary Inquiry (Bundestag) 2020–2021. Primary legislative investigation into BaFin conduct, audit failure, and regulatory response.

Criminal indictments. Munich prosecutor indictments of Markus Braun and others; ongoing proceedings.

Financial Times investigative record. Dan McCrum and colleagues; primary journalistic documentation from 2015-2020.

This case does not assess individual moral culpability, prosecutorial adequacy, or political accountability. Its purpose is to establish, on the public record, how narrative coherence, authority signalling, linguistic stabilisation, and regulatory deference combined to sustain a false reality until collapse was unavoidable.

CASE 5

THE DIALOGUE OF THE DEAF

A DIAGNOSTIC

Mapping the Post Office Horizon Conflict, Why Years of Communication Produced Zero Understanding

Purpose and scope. This case applies the Meaning Mesh Breakdown, introduced in Chapter 3, to the interaction patterns documented during the Post Office Horizon scandal. Unlike the forensic cases, which examine institutional failure at case level, This case operates at conversation level. It exists to show why years of communication between sub-postmasters and the Post Office produced zero shared understanding, not because people were not speaking, but because they were speaking from structurally incompatible definitions of what truth means. This is a diagnostic tool, not a historical record. The forensic record is in Case 1.

5.1 The Forensic Archetype

The following exchange is an archetype of thousands of interactions between sub-postmasters (SPMs) and the Horizon helpline or Post Office investigators between 2000 and 2015. It is reconstructed from the linguistic framing patterns documented in Horizon IT Inquiry transcripts. No single exchange is quoted verbatim. The pattern is uncontested.

SPM: 'I have a discrepancy of £4,000. It appeared instantly when the screen froze. I haven't taken a penny. This is a computer error.'

Helpline: 'Horizon is robust. We are not seeing errors elsewhere in the network. If the system shows a shortfall, you are contractually liable to make it good.'

SPM: 'But I can't pay money I don't have for a loss that isn't real. You need to check the logs.'

Investigator: 'The system record is the source of truth. Without evidence of theft by a third party, the transaction is your responsibility. Are you admitting to false accounting?'

This exchange did not fail because one party was dishonest. It failed because the two parties were operating from incompatible definitions of truth, definitions that made productive communication structurally impossible regardless of the goodwill, patience, or persistence of either side.

5.2 Mapping the Deadlock - The Meaning Mesh Breakdown

When we plot this interaction onto the Meaning Mesh Breakdown, we see why resolution was structurally impossible. The institution and the sub-postmaster were not simply disagreeing. They were occupying diagonally opposite quadrants with no shared protocol for moving between them.

INSTITUTIONAL SIDE	COUNTER SIDE
Quadrant 1 - Institutional Narrative **Authorised story. Validated internally.** **• Position: 'Horizon is robust.'** **• Logic: admitting a flaw threatens the integrity of the entire network's accounting**	Quadrant 2 - Counter-Narrative Lived signals. Alternative interpretations. • Position: 'I didn't take it. The screen froze.' • Problem: this narrative has no currency within the institution's framework; dismissed as anecdotal
Quadrant 3 - Assumptions Unspoken premises. **Never surfaced. Never tested.** **• Hidden belief: 'If a shortfall exists, someone must have caused it'** **• Structural blind spot; the assumption that the digital record is the physical reality**	Quadrant 4 - Evidence Test (The Missing Diagonal) The diagonal the institution refused to enter. • Missing data: REP-7F bug, TRA-00 rollback errors, raw transaction logs • The block: defence teams were routinely denied access to error logs

5.3 The Semantic Incompatibility

The Structural Deadlock occurs because the diagonal link between the Evidence Test quadrant and the Institutional Narrative quadrant is broken. In a functioning system, evidence updates the narrative: the institution enters Quadrant 4, examines the data, and revises its position in Quadrant 1. In the Horizon conflict, the institution refused to enter Quadrant 4. The sub-postmaster was attempting to move the conversation from Quadrant 3 (Assumptions) to Quadrant 4 (Evidence Test): 'Check the logs.' The institution blocked this move and forced the conversation back to Quadrant 1 (Narrative): 'You are contractually liable.' This is Semantic Incompatibility. The two parties held different definitions of truth: to the Post Office, truth meant consistency with the system, coherence with the institutional record. To the sub-postmaster, truth meant consistency with reality, correspondence with what physically happened. Because these definitions were incompatible, no amount of communication could resolve the dispute.

5.4 Why More Communication Made Things Worse

The Dialogue of the Deaf is not a failure of quantity of communication. Sub-postmasters communicated extensively, they called the helpline repeatedly, wrote letters, engaged solicitors, and appeared before courts. More communication did not produce more understanding. It produced more volume in a system that had no mechanism to process the signal those communications carried. The institution's response to volume was to categorise it, to route each new communication into an existing narrative category that confirmed what the institution already believed. The sub-postmaster who wrote a fifth letter received the same response as the sub-postmaster who wrote the first.

Not because the institution was not reading, but because its reading was performed through a framework that could only produce one conclusion.

5.5 The Missing Mechanism

For the deadlock to have been broken, one structural condition needed to exist: a mechanism through which evidence from Quadrant 4 could have interrupted execution in Quadrant 1. That mechanism would have required: access to the error logs (which was denied); authority for the sub-postmaster's account to be treated as a signal rather than as noise; a pathway through which 'this may be a system error' could become an institutional question rather than an institutional threat; and permission for the Evidence Test to update the Institutional Narrative. None of these existed. Not because they were deliberately removed, but because the institution had not been designed with the assumption that its system could be wrong. This is the design failure that the Horizon case makes visible: an institution does not need to be malicious to produce a Dialogue of the Deaf. It only needs to be designed in a way that makes self-correction structurally unavailable.

5.6 Application to AI-Mediated Decisions

The Horizon pattern now appears in a new form wherever AI systems produce outputs that are treated as authoritative and where the people most affected by those outputs lack a structural mechanism to challenge them. The welfare recipient who receives a Robodebt notice and cannot successfully dispute it.

The mortgage applicant whose AI-generated credit score contains an error they cannot access the reasoning to contest. The hiring candidate ranked below the threshold by an automated screening system with no explanation pathway. In each case, the same four-quadrant structure is present: an institutional narrative (the system output), a counter-narrative (the individual's account of their own circumstances), an assumption (the system is reliable), and a missing diagonal (the evidence that would test that assumption is inaccessible).

5.7 Conclusion: A Failure of Listening Architecture

The Meaning Mesh Breakdown is a diagnostic for this condition. An organisation that can map its own dispute resolution processes onto these four quadrants and identify a broken diagonal is an organisation that has located the structural site of its next Dialogue of the Deaf, before it becomes irreversible.

This diagnostic demonstrates that the failure was not primarily technical. The Horizon bugs were real and serious. But bugs alone did not produce sixteen years of wrongful prosecutions. What produced them was a design failure in the architecture of listening.

See also: Case 1 (Post Office Horizon Forensic Record) for the full evidential basis of the case documented here.

CASE 6

THE CHALLENGER

Space Shuttle Challenger, January 28, 1986, How Engineering Truth Was Demoted to Attitude, and Why the Signal That Could Have Stopped the Launch Was Never Heard

Scope and purpose. This case documents the publicly established forensic record of the Space Shuttle Challenger disaster of January 28, 1986. It exists to demonstrate, using findings from the Rogers Commission and Diane Vaughan's forensic study, how a valid engineering signal was categorised as an attitude problem rather than a strategic warning, and how schedule pressure, authority dynamics, and the normalisation of deviance combined to ensure that the signal could not interrupt execution. Nothing in This case advances a novel claim. All material is drawn from primary commission findings, engineering records, witness testimony, and the published forensic literature. Interpretation and implication are addressed in the main body of the book. This case is evidentiary only.

CASE SUMMARY	Space Shuttle Challenger - January 28, 1986
Mission	STS-51-L, Space Shuttle Challenger
Date	January 28, 1986
Outcome	Vehicle disintegrated 73 seconds after launch; all 7 crew members killed
Physical cause	O-ring failure in right solid rocket booster field joint; hot gas breach; structural failure
Structural cause	Interpretive Demotion: engineering concern recategorised as attitude problem rather than stopping signal
Key signal holder	Roger Boisjoly - Thiokol engineer; raised O-ring concerns formally in July 1985
Launch temperature	28°F (–18°C) at launch - significantly below any previous launch; below O-ring qualification range
Commission	Presidential Commission on the Space Shuttle Challenger Accident (Rogers Commission, 1986)
Chapter reference	Chapter 4 - The Human Signal; Chapter 8 - AI as Frozen Meaning (Vaughan extension)
Primary sources	Rogers Commission Report (1986); Vaughan, The Challenger Launch Decision (1996)

6.1 Engineering Context and Prior Warning

The solid rocket boosters used on Space Shuttle missions were sealed at their field joints by O-rings, rubber seals designed to prevent hot combustion gases from escaping the booster casing. The O-rings' resilience was temperature-dependent: at lower temperatures, the rubber became less pliable and less capable of sealing the joint effectively under the pressure of launch. This was not a post-disaster discovery. Roger Boisjoly, a senior engineer at Morton Thiokol, had been raising concerns about O-ring performance at low temperatures since at least 1985. In July 1985, he wrote a formal internal memorandum warning that O-ring erosion under cold conditions represented a potential catastrophic failure mode, stating explicitly that the condition could result in 'a catastrophe of the highest order, loss of human life.' The memo was received, filed, and did not interrupt the launch schedule.

6.2 The Night Before: The Teleconference

On the evening of January 27, 1986, engineers at Morton Thiokol participated in a teleconference with NASA managers. The forecast temperature at launch time was approximately 28°F (–18°C), significantly colder than any previous Shuttle launch, and well below the qualified temperature range for O-ring performance. Thiokol engineers presented data showing a correlation between low temperature and O-ring erosion. Their recommendation was unambiguous: do not launch below 53°F (12°C). The NASA managers' response is the forensic pivot point of the case. Lawrence Mulloy expressed frustration: 'My God, Thiokol, when do you want me to launch, next April?' George Hardy stated he was 'appalled' by the recommendation. The pressure was direct, documented, and exercised by the people with authority over the launch decision.

6.3 The Management Caucus and the Reversal

Thiokol management requested a caucus. During it, Vice President Joe Kilminster asked the engineers to take off their 'engineering hats' and put on their 'management hats.' This instruction is the most precise available articulation of Interpretive Demotion in the primary record. It did not ask engineers to review their data. It asked them to change the frame through which they were interpreting the data they already had. The engineering concern was not refuted. Its status was changed. Boisjoly and Thompson continued to argue against launch during the caucus. They were overruled. When the teleconference resumed, Kilminster recommended proceeding with the launch. The decision was made by managers, not engineers. The engineers who held the signal were present throughout. They were not heard.

6.4 Launch and Disintegration

> 07:38 EST - Space Shuttle Challenger launches from Kennedy Space Center. Air temperature: approximately 28°F (–18°C), the coldest launch in Shuttle history.
>
> T+0.678s - High-speed camera footage reveals a puff of grey smoke from the right solid rocket booster field joint, the first indication of O-ring seal failure.
>
> T+58.788s - Flame breakthrough from the right booster field joint becomes visible in telemetry.
>
> T+72.204s - Structural failure of the external tank. Vehicle aerodynamically disintegrates at approximately 46,000 feet. All seven crew members are killed: Francis Scobee, Michael Smith, Judith Resnik, Ellison Onizuka, Ronald McNair, Gregory Jarvis, and Christa McAuliffe.

The O-rings failed exactly as Boisjoly had predicted they would at the temperatures experienced on launch day.

6.5 The Rogers Commission Findings

The Rogers Commission concluded that the O-ring erosion problem had been known to NASA management prior to the launch, that concerns had been raised and documented, and that the decision to launch was made without full disclosure of the engineering objections to senior NASA management. It identified a flawed decision-making process in which information did not flow upward correctly, and an organisational culture in which schedule pressure took precedence over safety concerns.

Richard Feynman: 'For a successful technology, reality must take precedence over public relations, for nature cannot be fooled.'

6.6 Vaughan's Contribution: The Normalisation of Deviance

Diane Vaughan's forensic study, The Challenger Launch Decision (1996), identified the normalisation of deviancw; the gradual, collective renegotiation through which NASA and Thiokol had come to treat O-ring erosion as an acceptable risk rather than a danger signal. Across multiple previous flights, O-ring erosion had been observed. Each flight that returned safely became a data point that quietly revised the organisation's threshold for acceptable risk. The extension this book makes is precise: when normalised deviance is encoded into automated systems, the renegotiation mechanism disappears entirely. There is no meeting in which the threshold is quietly revised. There is no Boisjoly who can refuse. The system enforces what humans accommodated, without the possibility of a human sitting in a teleconference and saying 'I will not sign off on this.' The deviance does not continue to normalise. It institutionalises.

6.7 Mapping to the Predictable Volatility Framework

Meaning Mesh Analysis

PILLAR	DESIGN REQUIREMENT	WHAT THIS CASE EXHIBITED	STATUS
Definition Integrity	The definition of 'acceptable launch conditions' is explicit, shared identically across engineering and management layers, and grounded in the physical performance specifications of the hardware rather than in schedule requirements	Engineers defined 'acceptable' by O-ring performance data: below 53°F, the risk was unacceptable. Management defined 'acceptable' by precedent: previous launches had returned safely. These definitions were incompatible and the gap was never formally surfaced. Authority resolved the ambiguity by declaration rather than by evidence.	ABSENT
Context Preservation	Engineering concerns travel upward through the organisation retaining their full technical context, including the data, the methodology, and the confidence level of the assessment - so that decision authority can evaluate them on their merits	When Thiokol management reversed the engineering recommendation during the caucus, the engineers' full technical context was not transmitted to senior NASA management. The launch recommendation arrived stripped of the dissent that had preceded it.	ABSENT

Contestability	A formally protected pathway exists for engineers to maintain a safety objection against management pressure and have that objection reach a decision authority capable of acting on it, without professional consequence	Boisjoly and Thompson argued against launch throughout the caucus and were overruled. The instruction to 'put on management hats' was the explicit reclassification of an engineering concern as an attitude problem. No formal mechanism existed through which engineers could maintain their objection once management had made its decision. The Right to Stop did not exist.	**DEGRADED**
Drift Detection	Monitoring exists to detect when the organisation's threshold for acceptable risk has shifted, when the normalisation of deviance has moved the definition of 'safe' away from the original engineering specification	The normalisation of deviance Vaughan describes operated precisely through the absence of drift detection. Each safe return following O-ring erosion was treated as confirmation rather than as a warning. No mechanism existed to flag that the cumulative accommodation of erosion events was moving the organisation's risk threshold away from the hardware's design limits.	**ABSENT**

Decision Architecture Failure

The Challenger case illustrates a precise Decision Architecture failure: the people with the signal had no authority, and the people with authority did not have access to the signal in its full form. The architecture through which information moved between these two groups was designed for alignment rather than contestation. When alignment required suppressing the signal, the architecture suppressed it.

The Predictable Volatility Pattern

Signals were present, documented, and named. Boisjoly's July 1985 memo named the catastrophic failure mode explicitly. The night-before teleconference made the engineering position unambiguous.

Interpretive Demotion was direct and documented. The instruction to 'put on management hats' is the clearest available primary record of the mechanism this book describes.

Decision Architecture enforced the wrong outcome. The process ran correctly. Engineers were consulted. A recommendation was made. But the architecture was designed for consensus rather than contestation.

Normalisation of deviance removed the threshold. Prior flights with O-ring erosion that returned safely had progressively moved the organisation's definition of acceptable risk away from the hardware's design limits.

The Right to Stop did not exist. Boisjoly could object. He could not stop. That distinction killed seven people.

6.8 Primary Sources and Evidential Basis

Rogers Commission Report (1986). Primary forensic record of the failure, the decision-making process, and the commission's findings.

Vaughan, D. The Challenger Launch Decision (1996). University of Chicago Press. The definitive sociological forensic analysis of the normalisation of deviance.

This case does not assess individual moral responsibility or the adequacy of NASA's subsequent reforms. Its purpose is to establish, on the public record, how Interpretive Demotion combined with the normalisation of deviance and the absence of a formal Right to Stop to produce a launch decision that the people with the most accurate information could not prevent.

CASE 7

VOLKSWAGEN DIESEL EMISSIONS

Authority vs Physics, How Organisational Commitment Overrode Engineering Reality, and Software Was Used to Reconcile the Irreconcilable

Scope and purpose. This case documents the publicly established forensic record of the Volkswagen diesel emissions scandal. It exists to demonstrate, using uncontested findings, how organisational authority can override physical and engineering constraints, and how language, hierarchy, and incentive structures can convert technical impossibility into institutional compliance. Nothing in This case advances a novel allegation. All material is drawn from court filings and criminal proceedings, regulatory enforcement actions and settlements, and internal disclosures and testimony. Interpretation and implication are addressed in the main body of the book. This case is evidentiary only.

CASE SUMMARY	Volkswagen Diesel Emissions Scandal – 2007-2015
Organisation	Volkswagen AG; including subsidiaries Audi, SEAT, Škoda, Porsche
Strategy	Clean Diesel - U.S. market expansion via premium diesel performance claims
Fraud period	Approximately 2007-2015 (defeat device active)
Vehicles affected	Approximately 11 million globally; approximately 600,000 in the U.S.
Regulatory trigger	U.S. EPA Notice of Violation, September 2015
Mechanism	Defeat device, software detecting test conditions and altering emissions output
Penalties	Over $30 billion in global settlements, fines, buybacks, and compensation
Criminal proceedings	U.S. guilty plea (VW entity); individual convictions including senior executives
Chapter reference	Chapter 5 - Authority meets Interpretation
Primary sources	U.S. DOJ settlement documents; EPA enforcement record; Braunschweig criminal proceedings

7.1 Strategic Context and Organisational Commitments

In the early 2000s, Volkswagen AG articulated a strategic objective of becoming the world's leading automobile manufacturer by volume. A central component was significant expansion in the United States market, where emissions regulations were among the most stringent globally. Volkswagen's proposed solution was branded as Clean Diesel: a diesel engine platform intended to deliver simultaneously high engine performance, strong fuel efficiency, and compliance with U.S. nitrogen oxide (NOx) emissions standards. These three objectives were treated internally and externally as non-negotiable and mutually achievable.

7.2 Engineering Assessment and Physical Constraints

By approximately 2006-2007, Volkswagen engineering teams reached a consistent technical assessment: given available emissions control technology, vehicle architecture constraints, and cost and pricing targets, it was not possible to achieve U.S. NOx emissions compliance, required performance characteristics, and competitive pricing simultaneously. This assessment was not isolated, speculative, or unresolved. It reflected physical and chemical limits understood across the automotive industry. Internal discussions acknowledged that emissions-compliant solutions would either reduce engine performance or materially increase cost. Both outcomes conflicted with strategic commitments already communicated by senior leadership. The engineering reality was clear and documented. The question was what the organisation would do when authority met physics and physics did not yield.

7.3 Authority Response and the Defeat Device

Rather than revisiting the underlying strategic assumptions, leadership reaffirmed the Clean Diesel objective. Engineering constraints were reframed as delivery challenges. The governing question shifted decisively from 'Is this physically achievable?' to 'How will this be delivered?' This reframing is the pivotal moment in the Volkswagen failure, the point at which interpretation was placed downstream of authority. To reconcile the conflict between authority and physics, software was introduced into the emissions control system. This software detected when a vehicle was undergoing regulatory testing and altered engine behaviour to ensure compliance under test conditions. During normal driving, emissions controls were reduced or disabled, and performance and fuel economy targets were preserved.

7.4 Knowledge Distribution and the Accumulation of Meaning Debt

Subsequent investigations established that multiple engineering and management teams understood the software's functional behaviour. Engineers operated in an environment where raising feasibility objections threatened career progression, ambiguity was safer than explicit refusal, and delivery aligned with incentive structures. By enforcing the meaning 'compliant Clean Diesel' despite physical contradiction, Volkswagen accumulated Meaning Debt across each model year, each market, and each regulatory submission. Each vehicle produced extended the exposure. Each year of sales expanded the population of vehicles requiring remediation. Success delayed correction.

7.5 Regulatory Findings and Legal Outcomes

FINDING	OUTCOME
U.S. criminal plea	Volkswagen entity pleaded guilty to conspiracy, obstruction, and wire fraud (2017)
U.S. civil settlements	Approximately $15 billion in consumer compensation, buybacks, and environmental remediation (2016)
Global penalties	Over $30 billion in total fines, settlements, and remediation across jurisdictions
Individual convictions	Senior VW executives convicted in U.S. proceedings; German prosecutions ongoing
German criminal findings	Braunschweig proceedings confirm systematic knowledge within senior management

7.6 Mapping to the Predictable Volatility Framework

Meaning Mesh Analysis

PILLAR	DESIGN REQUIREMENT	WHAT THIS CASE EXHIBITED	STATUS
Definition Integrity	The definition of 'compliance' is explicit, shared across engineering, management, regulatory submissions, and end consumers, and means the same thing in testing and in operation	'Compliant' meant 'compliant in testing' internally and 'compliant in use' to every regulator and consumer. This definitional gap was the engine of the fraud, maintained deliberately through linguistic abstraction across every communication channel.	**ABSENT**
Context Preservation	Performance claims carry the context of the conditions under which they are valid, including the specific test conditions that differ from real-world operation	Emissions compliance claims were valid only under test conditions. This context was stripped from every regulatory submission, consumer communication, and sales claim. Context was the omission that made the fraud technically defensible at each step.	**ABSENT**

Contestability	A structured, protected mechanism exists for engineers to surface the finding that a strategic objective is physically unachievable and have that finding reach decision authority intact	The mechanism that should have produced coherence, escalation, existed structurally but was rendered functionally useless by incentive alignment and authority pressure. Engineers escalated into silence. 'This cannot be done' had no pathway to reach strategic decision-making intact.	DEGRADED
Drift Detection	Monitoring exists to detect when the gap between claimed compliance and actual emissions performance is growing, not just whether the system passes its own test criteria	No internal monitoring mechanism was designed to detect the growing gap between test performance and real-world emissions. The defeat device was itself the mechanism that concealed drift. Each passing test confirmed apparent compliance while actual emissions diverged further.	ABSENT

Decision Architecture Failure

The Volkswagen case illustrates the specific failure mode the PV framework calls authority without interpretive accountability; senior leadership set objectives that were physically unachievable and structured the organisation such that the finding of impossibility could not travel upward intact.

The Predictable Volatility Pattern

Signals were present. Engineers documented the constraint. Internal assessments established physical impossibility. The signal was clear.

The Meaning Mesh was closed at the authority layer. Strategic commitment became organisational reality regardless of technical finding.

Decision Architecture enforced compliance, not truth. The structure rewarded delivery and penalised objection. Those with information had no structural safety in sharing it.

Software mediated the irreconcilable. When authority and physics could not be made consistent, technology was used to make them appear consistent under controlled conditions.

Scale made correction catastrophic. The more successfully Clean Diesel sold, the larger the remediation liability. The cost of correction grew to a scale that required a decade of structured settlement.

7.7 Primary Sources and Evidential Basis

U.S. DOJ settlement documents 2016–2017. Guilty plea and civil settlement terms establishing the factual basis of the fraud.

EPA Notice of Violation. September 2015. Primary regulatory trigger document.

German criminal proceedings. Braunschweig court proceedings documenting senior management knowledge and participation.

This case does not assess moral character, cultural values, or ethical intent. Its purpose is to establish, on the public record, how authority, incentive alignment, language abstraction, and software mediation combined to enforce an interpretation that physical reality could not sustain.

CASE 8

THE COLUMBIA

Space Shuttle Columbia, February 1, 2003, How a Presumption of Futility Became a Decision Not to Know, and Why the Signal Was Demoted Before the Evidence Was Gathered

Scope and purpose. This case documents the publicly established forensic record of the Space Shuttle Columbia disaster of February 1, 2003. It exists to demonstrate, using findings from the Columbia Accident Investigation Board, how Interpretive Demotion can be driven not by commercial or schedule pressure but by a presumption of futility, a pre-emptive judgment that acting on a signal is impossible, made before the evidence to evaluate that judgment was gathered. Nothing in This case advances a novel claim. All material is drawn from the CAIB Report (2003), NASA engineering records, and disclosed internal communications. Interpretation and implication are addressed in the main body of the book. This case is evidentiary only.

CASE SUMMARY	Space Shuttle Columbia - February 1, 2003
Mission	STS-107, Space Shuttle Columbia
Launch date	January 16, 2003
Disaster date	February 1, 2003 (during re-entry)
Outcome	Vehicle disintegrated during re-entry; all 7 crew members killed
Physical cause	Foam strike to left wing leading edge on launch; breach of thermal protection system
Structural cause	Interpretive Demotion driven by presumption of futility: imaging requests denied on grounds that nothing could be done even if damage was confirmed
Key signal holders	Boeing engineers; NASA Debris Assessment Team
Imaging requests	Three formal requests for satellite imaging made; all denied or withdrawn
Investigation	Columbia Accident Investigation Board (CAIB) Report, August 2003
Chapter reference	Chapter 5 – Authority meets Interpretation
Primary sources	CAIB Report (2003); NASA engineering records and internal communications

8.1 The Foam Strike and Its Identification

On January 16, 2003, Space Shuttle Columbia launched from Kennedy Space Center on mission STS-107. Eighty-one seconds after launch, a piece of insulating foam broke away from the external tank's bipod ramp and struck the leading edge of the orbiter's left wing. The strike was captured on launch camera footage. During post-launch imagery review, engineers at Boeing identified the impact as larger and more direct than anything previously recorded in the Shuttle programme. The foam piece was estimated to be approximately 1.67 pounds, significantly heavier than foam strikes in prior missions, and it struck the wing's leading edge at approximately 545 miles per hour relative velocity. The thermal protection system tiles on the wing's leading edge were the primary defence against the extreme heat of re-entry. A breach in that system would allow superheated atmospheric gases to enter the wing structure. Engineers who understood this recognised that the foam strike warranted immediate assessment of potential damage.

8.2 The Imaging Requests and Their Denial

Flight Day 2-3 - Boeing engineers formally request Department of Defense satellite imaging of Columbia's left wing to assess potential thermal protection system damage.

Flight Day 3-4 - NASA mission management at Johnson Space Center declines to pursue the imaging request. The stated reasoning: even if damage is confirmed, nothing can be done about it. The orbiter is in orbit. There is no repair capability.

Flight Day 6 - A second request for imaging is made and again declined on similar grounds. A third request is initiated but withdrawn before formal submission, following the implicit message that further requests will not be productive.

February 1, 2003 - Columbia begins re-entry. Hot atmospheric gases enter the left wing through the foam strike breach. The wing structure fails. The vehicle disintegrates over Texas and Louisiana. All seven crew members are killed: Rick Husband, William McCool, Michael Anderson, Ilan Ramon, Kalpana Chawla, David Brown, and Laurel Clark.

8.3 The Presumption of Futility

The reasoning given for denying the imaging requests was not that the engineering concern was unfounded. It was that even if damage existed, nothing could be done about it. This reasoning, the presumption of futility, is the structural mechanism that distinguishes the Columbia case from Challenger. At Challenger, the engineering signal was demoted because it threatened a schedule commitment. At Columbia, the engineering signal was demoted because someone upstream had decided that acting on it was impossible: confirming the damage would serve no operational purpose if nothing could be done. The judgment that nothing could be done was made before the evidence was examined. Post-accident analysis by the CAIB determined that the damage was severe enough to be potentially catastrophic, and that had the imaging been obtained and the damage confirmed, emergency options would have been available, including extending the mission to allow time for a rescue attempt using Space Shuttle Atlantis, which was being prepared for a scheduled mission.

The presumption of futility was wrong. The judgment that 'nothing could be done' was reached without the information that would have tested it. And it was reached before that information was sought.

8.4 The CAIB Findings on Organisational Cause

The CAIB concluded: 'The Columbia accident is, in the Board's view, as much a failure of NASA's organisational culture as it is a hardware failure.' The Board found that NASA had developed a cultural trait of 'broken safety culture' in which schedule pressures and resource constraints had led the organisation to accept risks that should have prompted corrective action. It identified a pattern in which anomalies were normalised rather than investigated, a direct parallel to the normalisation of deviance Vaughan had identified in the Challenger analysis.

8.5 Columbia and Challenger Compared: Two Forms of Interpretive Demotion

At Challenger, the demotion was driven by external pressure, NASA management pushed back explicitly. At Columbia, the demotion was self-generated, mission management pre-emptively decided that the signal did not warrant a response, without explicit pushback from above. This self-generated form is in some respects more dangerous. At Challenger, there was a visible confrontation. At Columbia, the demotion occurred silently, through the routine operation of an information management system designed to absorb anomalies rather than escalate them.

8.6 Mapping to the Predictable Volatility Framework

Meaning Mesh Analysis

PILLAR	DESIGN REQUIREMENT	WHAT THIS CASE EXHIBITED	STATUS
Definition Integrity	The definition of 'acceptable mission risk' is grounded in explicit engineering specification, shared identically across the engineering teams who assess risk and the management teams who authorise continuation	Engineers defined the foam strike as a potential catastrophic risk requiring imaging to assess. Mission management defined it as a known-category anomaly consistent with prior flight experience and therefore not requiring escalation. These definitions were incompatible and the gap was resolved by authority rather than by evidence.	**ABSENT**

Context Preservation	Engineering concerns travel upward through the organisation retaining their full technical context, including the reasoning, the data, and the request for additional information, so that decision authority can evaluate them	The foam strike concern was communicated upward through a chain that progressively reduced its urgency. By the time it reached mission management, it had been contextualised as a known risk category rather than as an unresolved question requiring imaging to answer.	**ABSENT**
Contestability	Engineers who have identified a potential catastrophic risk can maintain their concern against a management determination that the risk is acceptable, with access to a decision authority capable of overriding mission management if the concern warrants it	After the second imaging request was denied, the engineering teams had no formal mechanism to escalate their concern to a level of authority above mission management. The third request was withdrawn before submission. The structure provided no pathway through which 'we need the imaging' could persist as an institutional requirement rather than a preference.	**DEGRADED**

Drift Detection	Monitoring exists to detect when the organisation's risk assessment practices have normalised anomalies that should trigger escalation, when foam strikes have been absorbed into routine risk management in a way that has moved the threshold away from the hardware's actual vulnerability	The normalisation of foam strikes across prior missions had moved the organisation's risk threshold such that a strike significantly larger than any previously recorded was processed through the same routine framework as minor strikes. No mechanism existed to detect that the current strike was categorically different from the incidents it was being compared to.	**ABSENT**

Decision Architecture Failure

The Columbia case illustrates a specific and particularly insidious form of Decision Architecture failure: the pre-emptive closure of inquiry. At Challenger, the Decision Architecture failed because it allowed authority to override a signal that had reached it. At Columbia, the Decision Architecture failed because it prevented the signal from generating the evidence that would have allowed it to be properly evaluated at all. The decision not to obtain imaging was itself a decision, one that determined the scope of information available to every subsequent decision in the mission.

The Predictable Volatility Pattern

Signals were present and specific. The foam strike was identified in launch footage. Engineers named it as larger and more concerning than any prior recorded strike. The imaging request was explicit: we need to see the wing.

Interpretive Demotion was driven by presumption of futility. The signal was not disputed. It was pre-emptively classified as not requiring a response on the grounds that nothing could be done - a conclusion reached before the evidence that would have tested it was examined.

Decision Architecture prevented the question from being answered. The denial of imaging requests closed off the evidence pathway. The engineers could not answer 'how serious is the damage?' because the mechanism for answering it had been shut down.

The presumption of futility was wrong. Post-accident analysis confirmed that rescue options existed. The judgment that nothing could be done was made without the information that would have revealed that something could.

Normalisation of prior anomalies set the frame. The history of foam strikes that had not caused catastrophic failure created the interpretive frame within which this strike was evaluated. That frame made the current strike look like a known risk category rather than an unresolved question.

8.7 Primary Sources and Evidential Basis

Columbia Accident Investigation Board Report (2003). August 2003. Primary forensic record of the physical cause, the organisational cause, and the Board's findings and recommendations.

NASA engineering records and internal communications. Disclosed during the CAIB investigation: Debris Assessment Team analyses, imaging request documentation, and mission management communications.

This case does not assess individual moral responsibility or the adequacy of NASA's subsequent reforms. Its purpose is to establish, on the public record, how the presumption of futility combined with the normalisation of prior anomalies and the absence of a formal evidence-gathering obligation to produce a disaster that the evidence, had it been sought, might have prevented.

See also: Case 6 (Space Shuttle Challenger, 1986), the paired case demonstrating two distinct forms of Interpretive Demotion from a single structural mechanism.

CASE 9

THE BOEING 737 MAX

Boeing 737 MAX / MCAS, October 2018 - March 2019 - How Four Incompatible Definitions of the Same System Produced 346 Deaths, and Why the Architecture Prevented Any Individual From Stopping It

Scope and purpose. This case documents the publicly established forensic record of the Boeing 737 MAX disasters of October 2018 and March 2019. It exists to demonstrate, using findings from the U.S. House Committee on Transportation and Infrastructure, the Joint Authorities Technical Review, the NTSB, and the subsequent criminal and civil proceedings, how four incompatible definitions of the MCAS system were simultaneously held by engineers, pilots, regulators, and the system itself, and how the architecture through which those definitions interacted prevented correction before 346 people died on two aircraft. Nothing in This case advances a novel claim. All material is drawn from primary investigation findings, engineering records, internal communications, and congressional testimony. Interpretation and implication are addressed in the main body of the book. This case is evidentiary only.

CASE SUMMARY	Boeing 737 MAX / MCAS Disasters
Flights	Lion Air Flight 610 - Jakarta, October 29, 2018 Ethiopian Airlines Flight 302 - Addis Ababa, March 10, 2019
Deaths	346 people killed across both crashes
System	MCAS - Maneuvering Characteristics Augmentation System
Physical cause	Erroneous angle-of-attack sensor input triggered MCAS; repeated nose-down commands overrode pilot inputs; crews were not informed MCAS existed
Structural cause	Managed Knowing: four incompatible definitions of MCAS held simultaneously by engineers, pilots, regulators and the aircraft; no mechanism to surface the divergence before deployment at scale
Grounding	737 MAX grounded globally March 13, 2019 - 20 months, the longest grounding of a commercial aircraft type in history
Financial cost	Estimated $20 billion in direct costs to Boeing; $2.5 billion criminal penalty
Chapter reference	Chapter 6 – Truth Rejection; Chapter 10 - The Design Obligation
Primary sources	U.S. House Committee Report (Sept 2020); JATR Report (Oct 2019); NTSB Safety Recommendation Report (Sept 2019); DOJ Criminal Information (Jan 2021)

9.1 The System and Its Development

The Boeing 737 MAX was developed as a fuel-efficient upgrade to the existing 737 NG series. A fundamental challenge in the programme was the placement of larger, more fuel-efficient engines. The CFM LEAP-1B engines were larger in diameter than those used on previous 737 variants and had to be mounted further forward and higher on the wing to maintain ground clearance. This changed the aircraft's aerodynamic characteristics, specifically, its handling at high angles of attack. At high angles of attack, the new engine position created a tendency for the nose to pitch upward more aggressively than on previous 737 variants, a handling difference that was commercially significant because Boeing had marketed the MAX to airlines on the basis that it could be flown by 737-rated pilots without expensive simulator retraining. MCAS - the Maneuvering Characteristics Augmentation System, was developed to address this handling difference by automatically pushing the nose down when the system detected a high angle-of-attack condition. The critical design decision that produced the subsequent disaster was that MCAS operated from data provided by a single angle-of-attack sensor, with no built-in redundancy, no crew alerting system, no published procedure for pilots to follow if MCAS activated incorrectly, and no reference in the flight manual.

9.2 Four Definitions of the Same System

The forensic record establishes that at the time the 737 MAX entered service, the people responsible for the aircraft held four fundamentally incompatible definitions of what MCAS was, and that none of those definitions corresponded to the system's actual operational behaviour.

ACTOR	DEFINITION OF MCAS
Boeing's engineers	A background stability-assist mechanism, designed to activate only in rare, specific aerodynamic conditions during manual flight.
Pilots	Nothing. MCAS was not included in the 737 MAX flight manual. Pilots had no knowledge that a system capable of overriding their inputs existed.
Regulators (FAA)	An extension of an existing trim system, not a fundamentally new behaviour requiring recertification or independent safety review.
In reality	A system that could repeatedly and aggressively override pilot input based on data from a single sensor, with no automatic cross-check, no crew alerting system, and no manual override procedure that pilots knew to apply.

The gap between the fourth row and the first three is the structural failure of the case. Every formal governance mechanism, the safety classification, the certification process, the flight manual, the training programme, operated on the first three definitions. None of them required anyone to act on the fourth.

Truth was not absent. It was distributed across four incompatible meanings, and no mechanism existed to surface the divergence before it became fatal.

9.3 Safety Classification and the Regulatory Failure

Boeing's initial safety analysis classified MCAS as a hazard level of 'major', a designation meaning the system's failure could cause serious injury but was not catastrophic. This classification determined the level of regulatory scrutiny applied to the system: a 'catastrophic' classification would have required independent FAA validation, while a 'major' classification allowed Boeing to self-certify under its Organisation Designation Authorisation. The initial 'major' classification was made when MCAS had a limited authority of

0.6 degrees per activation. During flight testing, Boeing engineers determined that MCAS needed significantly greater authority, and expanded it to 2.5 degrees per activation, a fourfold increase, removing the activation limit so MCAS could activate repeatedly without limit. This expansion fundamentally changed what MCAS was: a trim-assist became a primary flight control system capable of producing an unrecoverable dive. The safety classification was not updated. The process that had determined the hazard level was not repeated. The FAA certification was granted, on the basis of the original, limited-authority version of the system. The aircraft that entered service was not the aircraft that had been classified.

9.4 Lion Air Flight 610 - October 29, 2018

On October 29, 2018, Lion Air Flight 610 departed Jakarta with 189 people on board. Shortly after take-off, the left angle-of-attack sensor provided an erroneous reading. MCAS activated, pushing the nose down. The crew responded by pulling back on the control column, temporarily counteracting MCAS. MCAS deactivated and reactivated multiple times. The crew did not know MCAS existed. They had no procedure to follow. Flight 610 struck the Java Sea approximately thirteen minutes after take-off.

All 189 people on board were killed. Following the crash, Boeing issued an Operations Manual Bulletin describing MCAS and the procedure for managing an erroneous activation. The bulletin did not ground the aircraft, did not require simulator training, and described the procedure in technical language that presupposed pilots knew MCAS existed. The 737 MAX continued flying.

9.5 Ethiopian Airlines Flight 302 - March 10, 2019

On March 10, 2019, Ethiopian Airlines Flight 302 departed Addis Ababa with 157 people on board. Shortly after takeoff, an AoA sensor failure produced an erroneous reading. MCAS activated. The crew followed the Boeing runaway stabiliser procedure, cutting out the electric stabiliser trim. MCAS was disabled, but the stabiliser had been driven to an extreme position, and the aircraft was moving at high speed. Manual trim was physically impossible under the aerodynamic forces acting on the stabiliser. The crew re-engaged the electric trim to attempt recovery. MCAS reactivated. Flight 302 impacted the ground approximately six minutes after take-off. All 157 people on board were killed. The crew had followed the procedure they had been given. The procedure was insufficient for the conditions they faced. The 737 MAX was grounded by Chinese aviation authorities on March 11, 2019. The FAA grounded it on March 13th , two days later, and only after most other aviation authorities had already acted. It remained grounded for 20 months.

9.6 The Internal Record

During congressional and regulatory investigations, internal Boeing communications were disclosed that established the Managed Knowing condition with forensic precision. A 2016 internal message from a Boeing pilot referred to the 737 MAX simulator as 'running like a runaway train', predating the first crash by two years. Internal messages from 2018 showed Boeing employees discussing the pressure to avoid simulator training requirements for the MAX. A 2018 message between two Boeing pilots discussed MCAS activation characteristics; one pilot wrote 'I'll be honest with you, I'm worrying every time now', this exchange predated the Lion Air crash by several months. Documents showed that Boeing had considered adding a cockpit alert to warn pilots when the two AoA sensors disagreed, a feature that would have provided the Lion Air and Ethiopian crews with the information they needed. The feature was removed from the standard specification to reduce costs; airlines could purchase it as an optional add-on. Lion Air and Ethiopian Airlines had not done so.

The truth had been present for years. It had not been permitted to act.

9.7 Mapping to the Predictable Volatility Framework

Meaning Mesh Analysis

PILLAR	DESIGN REQUIREMENT	WHAT THIS CASE EXHIBITED	STATUS
Definition Integrity	The operational definition of MCAS, what it does, under what conditions, with what authority, and from what sensor inputs, is shared identically across the engineering, certification, training, and operational layers of the organisation	MCAS was defined differently by Boeing engineers (limited trim-assist), the FAA (extension of existing system), pilots (non-existent, not in the flight manual), and the aircraft's operational behaviour (aggressive primary flight control system with single-sensor dependency). The authority expansion from 0.6 to 2.5 degrees per activation was not reflected in the safety classification, the certification basis, or the pilot training programme.	**ABSENT**

Context Preservation	When the safety classification of MCAS is reviewed at each stage of the certification and operational process, it carries the full context of the system's actual behaviour, including its authority limits, its sensor inputs, and its failure modes	The safety classification was established for the original limited-authority version of MCAS and was not updated when authority was expanded. The context that would have required a 'catastrophic' classification, the expanded authority, the single-sensor dependency, the absence of crew alerting - was not carried forward through the certification process.	**ABSENT**
Contestability	A formal, protected pathway exists for engineers and safety personnel to maintain a concern about MCAS's operational characteristics against commercial and schedule pressure, and to have that concern reach a decision authority capable of acting on it	Internal communications establish that Boeing employees held concerns about MCAS that were not surfaced through the formal certification process. The internal culture around avoiding simulator training requirements created structural pressure against surfacing those concerns.	**DEGRADED**

Drift Detection	A mechanism exists to detect when the organisation's understanding of what MCAS does has drifted from its actual operational behaviour, specifically, when the authority expansion has created a system materially different from the one that was classified and certified	No mechanism existed to flag that the system certified by the FAA and the system operating on the aircraft were not the same system. The authority expansion was tracked in Boeing's engineering records but was not fed back into the certification process, the flight manual, or the pilot training programme.	**ABSENT**

Decision Architecture Failure

The Boeing case illustrates Decision Architecture failure at its most consequential. The people closest to consequence, airline pilots, were structurally excluded from the information they needed to recognise and respond to an MCAS failure. The people who held the fullest understanding of the system's actual behaviour, Boeing engineers, operated within an incentive architecture that rewarded avoiding the classification level that would have triggered additional scrutiny. The people with formal authority over the certification, the FAA, operated on the basis of information provided by the organisation they were certifying. Each layer of the architecture was performing its function. None of the functions included a requirement to surface the gap between the system's classified behaviour and its actual behaviour.

The Predictable Volatility Pattern

Signals were present, internal, and documented. Boeing employees identified concerns about MCAS's characteristics, about the pressure to avoid simulator training, and about the implications of the authority expansion. Those concerns existed in the internal record and did not interrupt the certification or deployment process.

Managed Knowing operated at institutional scale. Boeing knew MCAS existed and what it did. The FAA knew Boeing had an ODA framework under which self-certification occurred. Pilots did not know MCAS existed. Each actor's knowledge was accurate within its own frame. The aggregate knowledge, what the system actually was and how pilots would encounter it, was held by no single actor and surfaced through no single mechanism.

Interpretive Demotion was structural rather than interpersonal. No manager told an engineer to change their hat. Instead, the incentive architecture around avoiding simulator training requirements created structural pressure against surfacing the concerns that would have required a higher hazard classification. The demotion did not require a conversation. It required only the design of a certification framework that rewarded alignment over accuracy.

The Right to Stop was absent by design. The 346 people who died on Lion Air and Ethiopian Airlines were on aircraft operated by pilots who did not know MCAS existed, using procedures insufficient for the failure mode they encountered, flying a system classified as minor in a version that no longer corresponded to its operational reality.

9.8 The Criminal Resolution

In January 2021, Boeing entered into a deferred prosecution agreement with the U.S. Department of Justice. The agreement included a $2.5 billion resolution: $243.6 million in criminal fines, $1.77 billion in compensation to airlines, and $500 million to a fund for the families of victims. The DOJ's criminal information described Boeing's former Chief Technical Pilot as having conspired to defraud the FAA by making materially false statements about MCAS's operational characteristics during the certification process. Families of victims and aviation safety advocates criticised the agreement as insufficient. In 2024, the DOJ reopened its investigation following findings that Boeing had violated the terms of the agreement. As of publication, the legal proceedings remain active.

9.9 Primary Sources and Evidential Basis

U.S. House Committee on Transportation and Infrastructure - The Design, Development, and Certification of the Boeing 737 MAX. Final Committee Report, September 2020.

Joint Authorities Technical Review (JATR). Observations, Findings, and Recommendations to the Federal Aviation Administration. October 2019.

National Transportation Safety Board (NTSB). Assumptions Used in the Safety Assessment Process and the Effects of Multiple Alerts and Indications on Pilot Performance. Safety Recommendation Report, September 2019.

U.S. Department of Justice - Criminal Information and Deferred Prosecution Agreement. United States v. The Boeing Company. January 2021.

This case does not assess individual moral responsibility, the adequacy of Boeing's subsequent safety reforms, or the regulatory changes implemented by the FAA following the grounding. Its purpose is to establish, on the public record, how four incompatible definitions of the same system produced the conditions in which 346 people died on two aircraft within five months, and why no mechanism in the architecture, was capable of surfacing the divergence before that outcome occurred.

CASE 10

GLOBAL FINANCIAL CRISIS

Narrative Capture and Linguistic Engineering, How Language, Abstraction, and Institutional Incentives Combined to Suppress Judgment Until Collapse Was Unavoidable

Scope and purpose. This case documents the publicly established forensic record of how language, abstraction, and institutional incentives combined to suppress judgment in the period leading up to the Global Financial Crisis (2007–2009). It exists to demonstrate, using official findings and contemporaneous records, how narrative coherence replaced evaluation, allowing systemic risk to expand while appearing controlled. Nothing in This case advances a novel allegation. All material is drawn from the U.S. Financial Crisis Inquiry Commission reports, UK Parliamentary and regulatory inquiries, court judgments and settlement records, and contemporaneous regulatory and market disclosures. Interpretation and implication are addressed in the main body of the book. This case is evidentiary only.

CASE SUMMARY	Global Financial Crisis – 2007-2009
Period examined	Approximately 1998-2009 (expansion phase through collapse)
Primary failure period	2004–2007 (peak accumulation of Meaning Debt in the financial system)
Collapse trigger	U.S. housing price decline beginning 2006-2007; systemic exposure 2007-2008
Scale of impact	Global; estimated $22 trillion in household wealth destroyed in the U.S. alone
Key institutional actors	U.S. and global investment banks; credit rating agencies; regulators; institutional investors
Primary inquiry (U.S.)	Financial Crisis Inquiry Commission (FCIC) - Final Report, January 2011
Primary inquiry (UK)	Turner Review (FSA, 2009); Parliamentary Commission on Banking Standards (2013)
Structural mechanism	Abstraction ladder: mortgage → MBS → CDO → rated tranche → institutional portfolio
Chapter reference	Chapter 7 - When Narrative Replaces Judgement

10.1 The Abstraction Ladder - How Risk Became Invisible

At the point of origination, the underlying asset was concrete: a residential mortgage, issued to an identifiable borrower, with observable income, credit, and repayment capacity. As assets moved through the financial system, they were progressively abstracted through a series of transformations that each added institutional legitimacy and removed interpretive legibility.

ABSTRACTION LEVEL	WHAT WAS LOST
Level 1: Subprime mortgage	Individual borrower risk, visible, assessable, proximate to consequence
Level 2: Mortgage-backed security	Pooled cash flows, borrower identity removed; risk aggregated and redistributed
Level 3: Collateralised debt obligation	Structured redistribution of pooled risk into tranches, underlying loans two steps removed
Level 4: Senior tranche	Marketed as low risk due to structural payment priority, risk reframed as structure
Level 5: AAA-rated security	Regulatory and institutional permission encoded as a label, judgment outsourced to rating
Level 6: Portfolio holding	Institutional balance sheet entry, multiple abstractions from original borrower

At each step, distance from the borrower increased, interpretive burden shifted from judgment to structure, and responsibility migrated away from origination. Risk did not disappear. It was transformed into a form that required less human interpretation, and therefore generated less human resistance.

10.2 Ratings as Institutional Authority

Credit rating agencies occupied a structurally central role. Their ratings determined regulatory capital treatment, governed investment eligibility, and satisfied compliance requirements. A security rated AAA was treated as safe, not because its underlying loans were safe, but because the structure was deemed resilient. Subsequent disclosures confirmed that ratings assessed modelled loss distributions, not loan quality; correlations were assumed, not observed; and housing price declines were treated as implausible tail events. Judgment was not exercised at the point of use. It was deferred to the label.

10.3 Incentive Alignment Across Institutions

Narrative capture did not require coordination or conspiracy. It required aligned incentives across independent institutions, each responding rationally to its own structure. Investment banks earned fees on deal volume, incentive to originate and distribute, not to retain and evaluate. Rating agencies were compensated by issuers - incentive to rate competitively rather than conservatively. Regulators relied on ratings for capital adequacy, incentive to accept ratings rather than challenge them. Institutional investors relied on ratings for compliance, incentive to hold rated instruments rather than understand them. Originators sold mortgages immediately, incentive to originate, not to assess long-term repayment capacity. Challenging the abstraction imposed professional risk, reputational cost, and regulatory friction. Accepting it imposed none, until collapse.

10.4 Warnings Without Authority

Contrary to post-crisis simplifications, warnings existed. The FCIC documented concerns raised regarding underwriting standards, model assumptions, asset correlation, and liquidity dependence. These warnings were rarely suppressed directly. Concerns were contextualised as cyclical noise. Risk was reframed as model limitations. Dissent was deferred. Dissenters were isolated. Dissent survived. Authority did not accompany it.

10.5 Collapse and Meaning Debt Release

2007 Q2 - Bear Stearns hedge funds collapse; first visible systemic stress.

Aug 2007 - BNP Paribas freezes three investment funds - European market seizure begins.

Sep 2008 - Lehman Brothers files for bankruptcy - global credit markets freeze.

Oct 2008 - U.S. TARP ($700bn) enacted; global central bank coordinated action begins.

2008-09 - GDP contractions across U.S., UK, Eurozone; unemployment surge; sovereign debt stress.

10.6 Mapping to the Predictable Volatility Framework

Meaning Mesh Analysis

PILLAR	DESIGN REQUIREMENT	WHAT THIS CASE EXHIBITED	STATUS
Definition Integrity	Risk is named, defined, and understood consistently across originators, distributors, raters, regulators, and end investors, with shared understanding of what 'safe' means across the full chain	'Safe' meant AAA-rated to investors, structurally senior to arrangers, historically modelled to rating agencies, and capital-compliant to regulators. None of these definitions aligned with 'will repay principal in adverse conditions.' Meaning was consistent within each institution and inconsistent across the system.	**ABSENT**
Context Preservation	Risk assessments carry the economic assumptions, correlation estimates, and historical data constraints under which they were made, disclosed alongside the assessment so that end users can evaluate their basis	Every risk model embedded assumptions about housing prices, employment, and correlation that were never disclosed to end investors. The context - 'this model assumes housing prices do not decline nationally', was systematically stripped from every instrument sold. Models were presented as conclusions.	**ABSENT**

Contestability	When individual warnings conflict with the prevailing institutional narrative, a structured mechanism routes them to decision authority rather than absorbing them silently through contextualisation, deferral, or isolation	Warnings existed at every level of the system. None reached decision authority intact. Each was absorbed, reframed, contextualised, or deferred by the layer above. The incentive structure ensured that concerns were professionally costly to raise and institutionally costless to ignore.	**ABSENT**
Drift Detection	Monitoring tracks whether the meanings embedded in financial instruments, the correlation assumptions, the housing price expectations, the risk classifications, remain accurate as market conditions evolve	No mechanism existed to detect growing divergence between modelled risk and actual risk. Rising prices validated assumptions, confirming the narrative rather than testing it. Drift was invisible because the monitoring systems were calibrated against the models, not against reality.	**ABSENT**

Decision Architecture Failure

The GFC illustrates the failure mode the PV framework identifies as distributed responsibility without accountability nexus: every institution in the chain had a role, delegated judgment to the next institution, and bore no direct consequence for failing to verify at first principles. Responsibility circulated. It never landed.

The Predictable Volatility Pattern

Signals were abundant. Risk officers, external analysts, and investigative journalists all raised concerns. The signals existed across the system.

The Meaning Mesh was captured by narrative. The abstraction ladder progressively stripped interpretive content. By the time risk reached decision authority, it had been translated into a label that authorised confidence rather than caution.

Decision Architecture rewarded alignment. The incentive structure across every institution created rational motivation to accept the narrative and professional cost for challenging it.

Warnings were de-powered by structure. Dissent existed but lacked authority. The organisational architecture ensured that concerns were absorbed rather than elevated.

Scale made collapse systemic. What failed was not one mortgage, one bank, or one model. What failed was the shared meaning system that the entire financial architecture depended on. When it broke, it broke everywhere simultaneously.

10.7 Primary Sources and Evidential Basis

FCIC Final Report (2011). U.S. Financial Crisis Inquiry Commission. Primary U.S. government forensic record of systemic causes.

Turner Review (2009). UK Financial Services Authority. Primary UK regulatory analysis of structural failure.

Levin-Coburn Report (2011). U.S. Senate Permanent Subcommittee on Investigations. Detailed case studies of rating agency and bank conduct.

This case does not assess moral culpability, prosecutorial adequacy, or regulatory competence beyond documented findings. Its purpose is to establish, on the public record, how narrative capture, linguistic engineering, incentive alignment, and the structural suppression of dissent combined to produce a meaning system that the global financial architecture could not survive.

CASE 11

TOSHIBA CORPORATION

How Cultural Architecture Produced Interpretive Collapse, and Why Governance Without Contestability Is Theatre

Scope and purpose. This case is a forensic record, not an argument. It exists to demonstrate, through documented findings and uncontested evidence, how a structurally enforced cultural norm, the suppression of dissent in the name of organisational harmony, produced the same interpretive collapse patterns as algorithmic or technological failure. Nothing in This case advances a novel claim. All material is drawn from the Independent Investigation Committee report (2015), the Tokyo District Court ruling, the Financial Services Agency enforcement record, and published academic and regulatory analysis of Japanese corporate governance. Interpretation, analysis, and implication are addressed in the main body of the book. This case is evidentiary only.

CASE SUMMARY	Toshiba Corporation - Accounting Fraud 2008–2015
Organisation	Toshiba Corporation (Tokyo Stock Exchange: 6502)
Founded	1875; listed on Tokyo Stock Exchange since 1949
Fraud period	2008-2015 (seven years across three presidential tenures)
Scale of overstatement	¥152 billion (~$1.2 billion USD) in operating profit
Executives involved	Three successive presidents; 98 executives total
Exposure	Anonymous tip to SESC, January 2015
Investigation	Independent Investigation Committee report, July 2015
Regulatory action	Financial Services Agency enforcement; restatement required
Chapter reference	Chapter 7 - When Narrative Replaces Judgement
Primary source	Toshiba Independent Investigation Committee Report, 20 July 2015

11.1 The Cultural Mechanism: How Meaning Was Structurally Frozen

The 'Challenge' Mechanism

Japanese corporate discourse uses the term 'challenge' (chārenji) in a particular way. When a senior executive communicates a profit target as a 'challenge,' subordinates receive it as an unambiguous directive: this figure must be achieved. The word simultaneously signals impossibility of achievement by conventional means and permission, or instruction, to find other means. This mechanism is not corrupt in origin. It reflects a broader cultural norm around organisational harmony (wa) and the primacy of collective success. The problem is structural: it eliminates the space in which a subordinate can respond 'this cannot be achieved without compromising accuracy.' That response is not simply risky. It is culturally unspeakable.

Seven Years of Compounding

The fraud ran across three successive presidential tenures, each inheriting the adjusted figures as the baseline and compounding the overstatement. The system was not recalibrated at leadership transitions, the incoming president received the fraudulent baseline as institutional reality. The meaning had been frozen so completely that it survived the departure of the individuals who initiated it.

11.2 The Governance Paradox

Toshiba was an early voluntary adopter of a U.S.-style governance structure including a majority-independent board, a nomination committee, an audit committee, and a compensation committee with independent majority.

In formal terms, this was exemplary. Governance structures function on the assumption that the information flowing through them is interpretively intact. Independent directors can only challenge what they are shown. In Toshiba's case, each governance mechanism was operating on fraudulent inputs. The board received presentations prepared by management. The audit committee reviewed accounts that had been adjusted before they arrived.

Governance without contestability is theatre. The most sophisticated board structure becomes an ornament when the underlying meaning structure remains unchallengeable.

11.3 The Role of Whistleblowing

The fraud ended not because governance worked. It ended because someone stepped outside the structure entirely. In January 2015, an anonymous tip was submitted to Japan's Securities and Exchange Surveillance Commission. The tipster provided sufficient specificity for the SESC to initiate a formal enquiry, which triggered the Independent Investigation Committee. The committee's report confirmed the fraud in its entirety. This sequence is significant for the PV framework. Internal contestability had been structurally eliminated. The only mechanism that worked was external: a regulator receiving information that bypassed the organisation's interpretive layer altogether.

11.4 Consequences and Resolution

July 2015 - Independent Investigation Committee report published. Three presidents resign simultaneously.

September 2015 - Toshiba restates seven years of financial results. ¥152 billion overstatement confirmed.

2016 - Toshiba's nuclear division Westinghouse files for bankruptcy. Losses attributed partly to years of suppressed internal warnings about project overruns, a structurally related failure.

2021 - Third-party investigation finds Toshiba management colluded with the Japanese government to suppress shareholder votes at the 2020 AGM. A second structural governance failure.

December 2023 - Toshiba delisted from Tokyo Stock Exchange following a management buyout. Ends 74 years as a publicly listed company.

11.5 Mapping to the Predictable Volatility Framework

Meaning Mesh Analysis

PILLAR	DESIGN REQUIREMENT	WHAT THIS CASE EXHIBITED	STATUS
Definition Integrity	Performance targets have shared, explicit definitions understood identically by those who set them and those who report against them, with no ambiguity about whether targets represent aspirations or directives	The word 'challenge' was decoded differently by senders and receivers, senders understood aspiration, receivers understood instruction. The gap was never surfaced. No mechanism existed to make the definitional mismatch visible or to require its resolution.	**ABSENT**
Context Preservation	Financial figures travel upward through the organisation retaining the context of how they were produced, including whether targets were achieved by legitimate means or by accounting adjustment	Financial figures were adjusted before they travelled upward. The context, that targets were unachievable by legitimate means, was stripped in transit. Boards and committees received clean numbers that carried no record of their production method.	**ABSENT**

Contestability	Employees at any level can surface the finding that a target is unachievable without compromising accuracy, and have that finding reach decision authority through a protected pathway	Cultural architecture made questioning structurally impossible. Dissent was not suppressed through explicit instructions, it was suppressed through the absence of any pathway through which it could become organisationally legible.	**ABSENT**
Drift Detection	Monitoring exists to detect when reported financial performance is diverging from operational reality, flagging accumulating gaps before they compound across multiple reporting periods	No mechanism existed to detect the growing gap between reported and actual performance. The fraud compounded across seven years and three successive presidential tenures without the drift becoming visible through any internal monitoring channel. External detection required an anonymous tip to a regulator.	**ABSENT**

Decision Architecture Failure

The Toshiba case illustrates a specific failure of Decision Architecture: the concentration of interpretive authority in a hierarchy that had no obligation to pass signals upward accurately. In Toshiba, neither the structural capacity nor the cultural permission existed for those closest to information to challenge the meaning that senior leadership had assigned to it.

The Predictable Volatility Pattern

Signals were present. Employees knew the numbers were wrong. Concerns were raised. The signal existed.

The Meaning Mesh was compromised. The cultural architecture ensured that signals could not be converted into organisationally legible meaning. 'This number is inaccurate' had no pathway to become 'we must correct this number.'

Decision Architecture was structurally disabled. The governance apparatus operated on filtered inputs. Independent directors made legitimate decisions based on illegitimate information.

Feedback loops were severed. Because the information reaching decision-makers was fraudulent, corrective action could not be triggered. The feedback loop that should have detected and corrected the overstatement never activated.

11.6 The Universality Claim

The specific mechanism, the 'challenge' framing, the emphasis on organisational harmony, is culturally particular. The structural pattern is universal: wherever organisational culture suppresses the surfacing of dissonance, wherever hierarchy concentrates interpretive authority without providing accountability pathways, the same failure mode emerges. Toshiba is not a Japanese case study. It is a demonstration that interpretive collapse is a structural phenomenon.

11.7 Primary Sources and Evidential Basis

Independent Investigation Committee Report. Toshiba Corporation, 20 July 2015. Primary documentary record of the fraud, its mechanism, and the committee's findings.

Financial Services Agency enforcement record. Administrative sanction and penalty documentation, 2015.

This case does not assess moral character, cultural values, or ethical intent. Its purpose is to establish, on the public record, how cultural architecture, incentive alignment, and the structural suppression of dissent combined to produce interpretive collapse across seven years and three successive leadership tenures.

CASE 12

ZILLOW OFFERS

Frozen Meaning in AI Systems, How Probabilistic Insight Was Reassigned as Deterministic Command, and Why the System Could Not Change Its Mind in Time

Scope and purpose. This case documents the publicly established forensic record of Zillow Group's iBuying programme, Zillow Offers. It exists to demonstrate, using uncontested disclosures and outcomes, how probabilistic insight can be converted into deterministic command when embedded into automated systems, and how AI can freeze meaning at the precise moment when interpretation is most required. Nothing in This case advances a novel allegation. All material is drawn from Zillow Group SEC filings, investor communications and earnings calls, executive statements and post-mortem commentary, and contemporaneous financial and industry reporting. Interpretation and implication are addressed in the main body of the book. This case is evidentiary only.

CASE SUMMARY	Zillow Offers - Closure Q4 2021
Organisation	Zillow Group, Inc. (NASDAQ: Z/ZG)
Product	Zillow Offers - algorithmic iBuying programme
Launch	2018
Closure	Q4 2021 (programme halted; inventory liquidated)
Reported losses	Over $500 million in impairments and write-downs
Peak exposure	Approximately $3.8 billion in inventory at programme halt
Core algorithm	Zestimate - originally informational; reassigned as acquisition pricing authority
Failure mechanism	Frozen Meaning: probabilistic estimate embedded as deterministic command
Chapter reference	Chapter 8 – AI as Frozen Meaning
Primary sources	Zillow Group SEC filings; Q3 2021 earnings call; executive post-mortem statements

12.1 The Structural Reassignment of Meaning

The Zestimate was designed as an informational tool, a probabilistic estimate of property value, useful for orientation but not for operational decision-making. Zillow publicly disclosed median error rates in the range of 6–7%, and repeatedly emphasised that the Zestimate was not intended to be used as a definitive valuation. In 2018, Zillow launched Zillow Offers, an iBuying programme designed to algorithmically price homes, purchase them directly, renovate and resell them. This required a categorical shift in the role of the Zestimate. The original question, 'What is this home likely worth?', was replaced by 'What should we pay for this home?' The algorithm was not redesigned. The reassignment occurred at the organisational level: a probabilistic estimate was reassigned as a pricing authority. This is the structural moment the book calls Meaning Reassignment.

12.2 Automation and Organisational Embedding

As Zillow Offers scaled, algorithmic pricing outputs were embedded into acquisition decision pipelines, operational throughput targets, inventory management systems, and performance evaluation metrics. Once embedded, outputs were no longer advisory, pricing decisions became executable by default, and human teams were organised to operationalise results. The system no longer existed to inform judgment. Judgment existed to service the system.

12.3 Market Volatility and the Feedback Friction

During 2021, housing market conditions shifted rapidly. Local teams identified mismatches between algorithmic prices and observed market behaviour, but deviations were framed as execution or calibration issues, not as evidence that the core interpretive assumption was wrong. Overrides were institutionally discouraged. Correction mechanisms operated on timescales incompatible with market movement. The system continued to acquire inventory. Not because contradictory signals were absent. Because meaning had been fixed, and once meaning is fixed and embedded in automated execution, the cost of revision rises with every additional cycle of operation.

12.4 Humans in the Loop Without Interpretive Authority

Zillow repeatedly emphasised the presence of humans in the loop. In practice, humans adjusted parameters, handled exceptions, and managed operational stress. They did not possess authority to revise the system's core interpretive role. Once algorithmic pricing was tied to capital deployment and earnings expectations, challenging outputs carried organisational cost, slowing execution threatened guidance, and intervention became reputationally and politically risky.

Human involvement persisted. Judgment authority did not. This is the condition this book identifies as the most dangerous configuration in AI deployment: humans who can adjust but not revise, who can optimise but not stop, who are present in the loop but absent from the decision.

12.5 Post-Incident Characterisation and Outcome

By late 2021, the accumulated inventory represented a commitment that could not be unwound at scale without significant loss. Zillow halted new acquisitions, initiated large-scale inventory liquidation, and announced the shutdown of Zillow Offers. The programme resulted in reported losses exceeding $500 million. In post-mortem communications, Zillow characterised the failure as involving forecasting error, execution complexity, and operational challenges. These statements were not factually incorrect. They were incomplete. The failure was primarily one of semantic rigidity, the inability to revise meaning once automation scaled.

12.6 Mapping to the Predictable Volatility Framework

Meaning Mesh Analysis

PILLAR	DESIGN REQUIREMENT	WHAT THIS CASE EXHIBITED	STATUS
Definition Integrity	The meaning of model outputs, their confidence level, intended use, and limitations, is understood consistently across those who build them and those who act on them	The Zestimate's probabilistic nature was understood by data scientists and misaligned with the operational imperative of the iBuying programme. 'Estimate' and 'acquisition price' are not the same meaning, but the organisation embedded the estimate in a pipeline that treated it as the latter.	**ABSENT**
Context Preservation	The conditions under which the Zestimate was trained, including its error rate, its valid use cases, and the market conditions it assumes, are visible at the point of use and flagged when conditions change	The Zestimate was trained on historical data that did not include the volatility of 2021. Its training context was not attached to its outputs. When conditions shifted, the model continued to output numbers with no flag that its context had expired.	**ABSENT**

Contestability	When local signals conflict with model outputs, a structured escalation pathway exists to challenge the model's operational role, not merely to adjust its parameters	Local teams observed mismatches. Those observations had no clear escalation pathway to the model's operational authority. The structure ensured that ground-level signals could adjust parameters but could not challenge the model's fundamental role.	DEGRADED
Drift Detection	Monitoring exists to detect when the market conditions the Zestimate assumes are diverging from actual conditions, triggering review of the model's operational role before the divergence compounds	Losses were processed as signals to adjust buying behaviour, not to question the underlying assumption. The monitoring that existed was calibrated to optimise performance within the current framework, not to detect when the framework itself had become invalid.	ABSENT

Decision Architecture Failure

The Zillow Offers programme embedded the algorithm at the centre of its Decision Architecture without designing the interpretive revision layer that should surround every automated authority. The humans around the system had operational roles, not interpretive authority. This is the specific failure the PV framework calls the Missing Middle: the gap between the system's outputs and the organisational structure needed to question them.

The Predictable Volatility Pattern

Signals were present. Local market mismatches were visible. Teams on the ground identified them. The signal existed and was named.

The Meaning Mesh was frozen. Once the Zestimate became acquisition authority, no mechanism existed to revise that status as conditions changed. The tool became the decision. The decision could not be revised without dismantling the programme.

Decision Architecture lacked interpretive revision. Humans could adjust but not reassign. The model's role was structurally fixed.

Scale converted error into commitment. Each additional acquisition made the inventory position larger, the reversal more expensive, and the acknowledgment of failure more consequential.

AI froze meaning at scale. This is not a story about a bad algorithm. It is a story about an algorithm that performed as designed, in a role it was never designed for, at a scale that made course-correction progressively impossible.

12.7 Primary Sources and Evidential Basis

Zillow Group SEC filings. Q3 2021 10-Q and subsequent filings disclosing inventory write-downs and programme shutdown.

Q3 2021 earnings call. Executive statements characterising the failure and announcing programme closure.

Zestimate methodology disclosures. Zillow's public disclosures of median error rates and intended use limitations.

This case does not assess managerial competence, strategic intent, or regulatory compliance. Its purpose is to establish, on the public record, how the structural reassignment of a probabilistic tool as deterministic authority, combined with automation at scale and the absence of interpretive revision mechanisms, produced a failure that could not be corrected in time.

CASE 13

ROBODEBT

How Algorithmic Authority Replaced Human Interpretation and Produced Systematic Harm at Industrial Scale

Scope and purpose. This case documents the publicly established forensic record underlying Australia's Robodebt scheme and the subsequent Royal Commission findings. It exists to demonstrate, through primary evidence and uncontested findings, how an automated decision-making system came to function as the sole authorised source of meaning within an institution, producing systematic harm at industrial scale while human judgment was procedurally excluded. Nothing in This case advances a novel claim. All material is drawn from the Royal Commission into the Robodebt Scheme (Final Report, July 2023), Federal Court of Australia rulings, Senate Committee inquiry reports, class action settlement documentation, and Government disclosures and parliamentary records. Interpretation, analysis, and implication are addressed in the main body of the book. This case is evidentiary only.

CASE SUMMARY	Robodebt - Australia, 2016–2019
Programme	Online Compliance Intervention (OCI) - known as Robodebt
Jurisdiction	Australia - Department of Human Services (Services Australia)
Operational period	July 2016 - November 2019
Scale	Approximately 453,000 debt notices issued; $1.7 billion in alleged debts raised
Failure mechanism	Income averaging - annual ATO income divided by 26 to produce assumed fortnightly income
Legal finding	Federal Court (Amato v Commonwealth, 2020): income averaging debts unlawful
Royal Commission	Appointed 2022; Final Report July 2023
Net cost	$565 million to the Commonwealth - opposite of the savings intended
RC finding	'Venality, incompetence and cowardice' in programme administration
Chapter reference	Chapter 8 - AI as Frozen Meaning
Primary source	Royal Commission into the Robodebt Scheme - Final Report, July 2023

13.1 The Income Averaging Method and Its Mathematical Failure

At the core of the Robodebt system was a data-matching algorithm that compared fortnightly income reported by welfare recipients to Centrelink with annual income reported by employers to the Australian Taxation Office (ATO). The system's method - income averaging, divided annual ATO income by twenty-six fortnights to produce an assumed fortnightly income. This averaged figure was then compared to the recipient's Centrelink reports. The fundamental flaw was that income averaging assumes steady employment throughout the year. For anyone who worked irregularly, seasonally, or casually, this assumption produces false discrepancies, and in Australia's welfare population, irregular, seasonal, and casual work was the norm rather than the exception. The Federal Court ruled in Amato v Commonwealth (2020) that debts raised solely through income averaging were not valid debts under the Social Security Act. The method was unlawful. The programme had been running for three years.

13.2 The Reversal of Burden of Proof

Prior to Robodebt, the welfare agency was required to verify discrepancies before asserting a debt. Robodebt eliminated this verification step. The system's output became presumptive proof of debt. Recipients received automated notices stating the amount allegedly owed and that they had twenty-one days to provide evidence to the contrary. Many recipients could not respond effectively: payslips from years earlier had been discarded, employers had ceased trading, they lacked the digital literacy to navigate online dispute processes, and the notices did not explain how the debt had been calculated.

The burden of disproving the system's assertion fell entirely on individuals who were, in many cases, among the most administratively vulnerable in Australian society.

13.3 Internal Warnings and Structural Suppression

Pre-implementation warnings were raised by the Australian Council of Social Services. Frontline staff reported that the system was generating obviously incorrect debts at scale. Legal advice within government raised questions about the lawfulness of income averaging. The Royal Commission found that senior public servants did not disclose to Cabinet that the programme relied on income averaging, that legal advice questioning the lawfulness of the approach was not shared with decision-makers, and that the political and economic incentives to continue the programme overrode operational concerns. The warnings existed. The structure did not permit them to reach the decision-makers who could have acted on them.

13.4 Human Consequences and Documented Harms

Recipients reported anxiety, depression, and acute psychological distress following receipt of debt notices. The Commission heard from families of young people who died by suicide after receiving debt notices. Recipients had welfare payments garnished or tax refunds withheld in recovery of debts later found to be invalid. Some recipients entered financial hardship and credit damage from recovery actions for amounts they did not owe. Trust in government institutions was damaged across the population of 453,000 people directly affected.

13.5 Royal Commission Findings

FINDING	ROYAL COMMISSION CONCLUSION
On lawfulness	'The Scheme was unlawful.' Income averaging to raise debts without further verification was not authorised by the Social Security Act.
On ministerial responsibility	Ministers 'did not fulfil their obligation to ensure the program they administered was lawful.'
On public service conduct	Senior public servants 'misled Cabinet' by not disclosing the reliance on income averaging.
On institutional culture	'Venality, incompetence and cowardice' in the administration of the scheme.
On financial outcome	Net cost to the Commonwealth of $565 million, the opposite of the savings the programme was designed to achieve.

13.6 Mapping to the Predictable Volatility Framework

Meaning Mesh Analysis

PILLAR	DESIGN REQUIREMENT	WHAT THIS CASE EXHIBITED	STATUS
Definition Integrity	The definition of 'income' applied by the system matches the meaning of income as experienced by the recipients whose welfare entitlements it governs, accounting for the temporal patterns of irregular, seasonal, and casual work	The system's definition of 'income', annual ATO earnings averaged to fortnightly, did not match the reality of how welfare recipients earned money. The gap between institutional definition and lived reality produced 453,000 false assertions.	**ABSENT**
Context Preservation	The context in which income was earned, including its timing, seasonality, and relationship to welfare receipt, is preserved at the point of assessment and used to evaluate whether a discrepancy is genuine	The income averaging method stripped all temporal context from employment data. A person who worked intensively for three months and was unemployed for nine appeared identical to someone who worked steadily all year. Context, the single most important variable - was the datum the system eliminated.	**ABSENT**

Contestability	When system outputs conflict with recipient accounts, a structured mechanism exists to investigate the conflict and revise the system's output before debt is asserted, with the burden of proof remaining with the institution	The system's output was the assertion. There was no prior investigation. No mechanism existed to pause execution when evidence mounted that the outputs were systematically incorrect. Contestability was technically present but structurally impossible for the population most affected.	**DECORATIVE**
Drift Detection	Monitoring detects when the system's income averaging assumption is producing systematic errors - triggering review before the error compounds across hundreds of thousands of cases	No mechanism existed to detect the systematic pattern of false positives generated by the income averaging method. Escalating complaints were not analysed for systemic patterns. The programme continued issuing 20,000 notices per week at peak despite mounting evidence that the underlying method was producing systematically incorrect results.	**ABSENT**

Decision Architecture Failure

The Robodebt case illustrates the most complete form of Decision Architecture failure the PV framework identifies: the removal of human judgment from the decision chain entirely. Not the subordination of human judgment to algorithmic output, the procedural elimination of the step at which human judgment could have intervened. The architecture was not broken. It was designed to function without the check it needed.

The Predictable Volatility Pattern

Signals were present and escalating. Frontline staff reported incorrect debts. The Ombudsman received complaints. Legal advice questioned lawfulness. ACOSS warned before launch.

The Meaning Mesh was closed by design. The system's definition of income was embedded as authoritative. No pathway existed for lived reality to contest algorithmic assertion at the point of debt issuance.

Decision Architecture eliminated the human check. The verification step that had previously converted government data into legal debt was procedurally removed. The system's output became the decision.

Burden reversal compounded harm. By requiring recipients to disprove debts they had not incurred, the architecture placed the interpretive burden on those least equipped to carry it.

Political economy prevented correction. The programme's stated savings justified its continuation. Acknowledging systematic error required acknowledging that the savings were illusory.

13.7 Primary Sources and Evidential Basis

Royal Commission Final Report (July 2023). Primary documentary record of the scheme's operation, findings, and referrals.

Amato v Commonwealth (2020). Federal Court ruling establishing that income averaging debts were unlawful under the Social Security Act.

Class action settlement. Gordon Legal class action - $1.8 billion settlement (2021) - establishing the scale of unlawful debt recovery.

This case does not assess individual criminal liability beyond the Commission's referrals, or the adequacy of the subsequent political response. Its purpose is to establish, on the public record, how automated authority, burden reversal, and the structural suppression of dissent combined to produce 453,000 unlawful debt assertions and the documented suffering that followed.

CASE 14

TOESLAGENAFFAIRE

How Encoded Definitions, Automated Accusation, and the Absence of Contestation Produced Systematic Harm at Scale

Scope and purpose. This case documents the publicly established forensic record of the Dutch childcare benefits scandal, the Toeslagenaffaire, and the findings of the parliamentary inquiry, independent review commissions, and government investigations that followed. It exists to demonstrate, through primary evidence and uncontested findings, how an automated fraud detection system came to function as the sole authoritative interpreter of entitlement and compliance within a state institution - producing systematic harm at scale while the mechanisms through which families might have contested its outputs were procedurally absent or functionally inaccessible. Nothing in This case advances a novel claim. All material is drawn from the Parliamentary Inquiry Committee Report Ongekend Onrecht (December 2020), the Adviescommissie Uitvoering Toeslagen reports (the Van Dam Commission), Dutch Supreme Court and administrative court rulings, and government parliamentary correspondence and disclosures. Interpretation, analysis, and implication are addressed in the main body of the book. This case is evidentiary only.

CASE SUMMARY	Toeslagenaffaire - Netherlands, 2004–2019
Programme	Childcare benefit (kinderopvangtoeslag) administered by the Dutch Tax and Customs Administration (Belastingdienst Toeslagen)
Jurisdiction	Netherlands - Ministry of Finance / Belastingdienst
Operational period	Approx. 2004-2019 (systematic automated enforcement approx. 2013-2019)
Scale	Approximately 26,000 families wrongly designated as fraudsters; full repayment of all benefits demanded
Failure mechanism	Dual-nationality profiling combined with all-or-nothing benefit recovery logic - any administrative irregularity triggered full repayment demand regardless of materiality
Legal finding	Multiple administrative courts: recovery demands unlawful; proportionality requirements violated
Parliamentary inquiry	Committee Ongekend Onrecht (Unprecedented Injustice) - Final Report December 2020
Government outcome	Entire Dutch Cabinet (Rutte III) resigned January 2021 following inquiry findings
Chapter reference	Chapter 8 - AI as Frozen Meaning
Primary source	Parliamentary Inquiry Committee Report Ongekend Onrecht, December 2020

14.1 The Benefit System and Its Enforcement Architecture

The Dutch childcare benefit system required parents to meet eligibility conditions on an ongoing basis. Belastingdienst Toeslagen administered payments and conducted compliance checking. From approximately 2013, the system increasingly relied on automated risk-scoring and bulk recovery processes to identify suspected fraud and enforce repayment. The enforcement logic was binary: if a family was designated as having submitted an irregular application, including minor administrative errors, missing signatures, or documentation discrepancies that had no bearing on actual entitlement, the system demanded full repayment of all benefits received, sometimes spanning multiple years. There was no proportionality mechanism. A single-page error in a file covering three years of genuine entitlement produced the same outcome as deliberate fraud: total recovery demand.

The Parliamentary Inquiry Committee found that this all-or-nothing logic was not an operational error. It was a design choice, embedded in the enforcement architecture and sustained across multiple years despite mounting evidence of its consequences.

14.2 Ethnic and Dual-Nationality Profiling

Internal risk-scoring processes used nationality as a proxy variable for fraud risk. Families with dual nationality, including Dutch citizens with heritage in Turkey, Morocco, and other countries, were disproportionately flagged for investigation and recovery action. The Parliamentary Inquiry and subsequent government investigations confirmed that ethnicity and nationality had been used as selection criteria in automated screening processes. This was unlawful. The Dutch Data Protection Authority (Autoriteit Persoonsgegevens) confirmed the use of nationality as a risk indicator in its 2021 investigation.

The consequence was structural: the encoded definition of fraud risk was not constructed from behavioural indicators alone. It incorporated demographic characteristics, embedding a discriminatory proxy into the automated selection mechanism. Families who were accurately compliant with benefit rules were designated as suspects because of who they were, not what they had done.

14.3 The Absence of Contestation

Families who received repayment demands were formally entitled to appeal through administrative procedures. In practice, the contestation pathway was structurally inaccessible for the population most affected. Demands arrived without adequate explanation of the basis for designation. Internal objection procedures were handled by the same institution that issued the demand. Families who could not afford legal representation navigated a system designed by experts for experts.

Appeal processes were slow; in the interim, recovery actions continued, including garnishment of wages, tax refund interception, and referral to debt collection.

The Parliamentary Inquiry found that the Tax Administration had, on multiple documented occasions, withheld information from families and from courts that would have been material to appeal outcomes. Documents showing the basis for fraud designation were not disclosed. In cases where families obtained legal support and challenged demands, the institution did not provide the evidence base that would have allowed courts to assess the lawfulness of the automated determination. Contestation existed as a formal right. It was procedurally engineered out of practical reach.

14.4 Internal Warnings and Institutional Suppression

The Parliamentary Inquiry documented a sustained pattern of internal and external warnings that were received, registered, and not acted upon. These included: an internal memo from 2017 flagging the disproportionate impact of recovery processes on dual-nationality families; repeated warnings from the National Ombudsman between 2017 and 2019 that recovery demands were causing severe hardship and that procedural safeguards were inadequate; a 2019 report from an independent advisory body noting systemic problems with the application of the all-or-nothing recovery logic; and parliamentary questions from 2018 onwards that received responses which the Inquiry subsequently found to have been incomplete or misleading.

The Inquiry found that senior officials and ministers had been made aware of the systemic nature of the problem before it became public, and that the institutional response had been to manage disclosure rather than to correct the underlying mechanism.

14.5 Human Consequences and Documented Harms

The Inquiry documented that affected families experienced: complete loss of childcare access following benefit withdrawal; debt spirals driven by total repayment demands that exceeded annual household income; breakdown of marriages and family structures under financial pressure; and psychological harm including documented cases of depression, anxiety, and in a number of cases, suicide. Children were affected through forced removal from childcare, disruption to education, and family instability. The Inquiry described the harm as affecting the full arc of family life across the period of enforcement.

The scale, approximately 26,000 families, and the duration, over a decade, meant that the harm was not a collection of individual errors. It was a systemic outcome produced by an architecture that was functioning as designed.

14.6 Parliamentary Inquiry Findings

FINDING	PARLIAMENTARY INQUIRY CONCLUSION
On the enforcement architecture	The all-or-nothing recovery logic was disproportionate and produced unlawful outcomes at scale. It was a design choice, not an operational error.
On profiling	Nationality and ethnicity were used as fraud-risk proxies in automated screening. This was unlawful discrimination.
On contestation	The administrative appeal process was structurally inaccessible for affected families. The institution withheld material information from families and from courts.

On institutional awareness	Senior officials and ministers were aware of the systemic problem before it entered the public domain. The response was disclosure management, not correction.
On the report title	Ongekend Onrecht - Unprecedented Injustice. The Inquiry found the harm caused to families to be without precedent in Dutch post-war administrative history.
On government responsibility	The Cabinet accepted collective political responsibility and resigned in full on 15 January 2021.

14.7 Mapping to the Predictable Volatility Framework

Meaning Mesh Analysis

PILLAR	DESIGN REQUIREMENT	WHAT THIS CASE EXHIBITED	STATUS
Definition Integrity	The definition of 'fraud risk' applied by the system matches the actual pattern of fraudulent behaviour in the population it governs, accounting for administrative error, documentation complexity, and the characteristics of the eligible population	The system's definition of fraud risk incorporated nationality and dual citizenship as proxy variables. Administrative irregularities with no bearing on genuine entitlement produced identical outcomes to deliberate fraud. The gap between institutional definition and actual compliance produced 26,000 false fraud designations.	**ABSENT**
Context Preservation	The context in which administrative irregularities arose, including their nature, materiality, and relationship to actual entitlement, is preserved at the point of assessment and used to evaluate whether a discrepancy is genuine	The all-or-nothing recovery logic stripped all context from compliance assessment. A missing signature on a single document in a three-year claim history produced the same outcome as systematic non-compliance. Materiality, the single most important variable, was the datum the system eliminated.	**ABSENT**

Contestability	When system outputs conflict with applicant accounts, a structured mechanism exists to investigate the conflict and revise the system's output before recovery action proceeds, with the burden of proof remaining with the institution	Formal appeal rights existed. In practice, the institution withheld material information from families and courts, recovery actions continued during appeals, and the process was structurally inaccessible for the population most affected. Contestability was formally present and practically engineered out of reach.	**DECORATIVE**
Drift Detection	Monitoring detects when the risk-scoring and recovery system is producing systematic false positives, triggering review before the error compounds across tens of thousands of cases	No mechanism existed to detect the systematic pattern of unlawful designations. Internal warnings from 2017, Ombudsman findings, and administrative court rulings were received and not acted upon. The programme continued issuing recovery demands for years after evidence of systematic harm was documented.	**ABSENT**

Decision Architecture Failure

The Toeslagenaffaire illustrates a specific and consequential variant of Decision Architecture failure; the encoding of a discriminatory proxy into the automated selection mechanism, combined with the procedural removal of the contestation pathway through which that encoding could have been identified and corrected. The architecture was not broken. It was designed to function without the check it needed.

What distinguishes this case from Robodebt is not the structure of the failure, both involve automated accusation, burden reversal, and the suppression of correction, but the mechanism of discrimination. Robodebt encoded a mathematical assumption about how income is earned. Toeslagenaffaire encoded a demographic assumption about who is likely to commit fraud. In both cases, the encoding was institutionalised before execution began. In both cases, the means of correcting it were either absent or rendered inaccessible.

The Predictable Volatility Pattern

Signals were present and escalating. Internal memos flagged discriminatory patterns from 2017. The National Ombudsman issued repeated warnings. Administrative courts ruled recovery demands unlawful. Parliamentary questions were raised and answered incompletely.

The Meaning Mesh was closed by design. The system's definition of fraud risk was embedded as authoritative. No pathway existed for families to contest the basis of their designation at the point of accusation. Demographic characteristics were treated as evidence.

Decision Architecture eliminated the human check. The proportionality assessment that should have converted automated risk flags into individualised determinations was removed. The system's output was the decision.

Burden reversal compounded harm. By requiring families to disprove accusations they had not incurred, without access to the evidence base used to make them, the architecture placed the interpretive burden on those least equipped to carry it.

Institutional economy prevented correction. Acknowledging systematic error required acknowledging that the programme's fraud-prevention rationale was built on unlawful foundations. The institutional response was disclosure management.

14.8 Primary Sources and Evidential Basis

Parliamentary Inquiry Committee Report Ongekend Onrecht (December 2020). Primary documentary record of the scheme's operation, findings, and political consequences.

Adviescommissie Uitvoering Toeslagen - Van Dam Commission Reports (2020). Independent review of the Tax Administration's handling of benefit recovery and the treatment of affected families.

Autoriteit Persoonsgegevens Investigation Report (2021). Confirmation that nationality was used as a fraud-risk indicator in automated screening; finding of unlawful processing of personal data.

Dutch administrative court rulings (2019-2021). Multiple courts finding individual recovery demands disproportionate and unlawful; Supreme Court rulings establishing the legal standard the programme had failed to meet.

This case does not assess individual criminal liability, the adequacy of the compensation programme established from 2021, or the political consequences beyond the Cabinet resignation. Its purpose is to establish, on the public record, how encoded demographic profiling, the all-or-nothing enforcement architecture, and the structural inaccessibility of contestation combined to produce 26,000 unlawful fraud designations and the documented suffering that followed.

EXECUTIVE SUMMARY

THE PREDICTABLE VOLATILITY FRAMEWORK

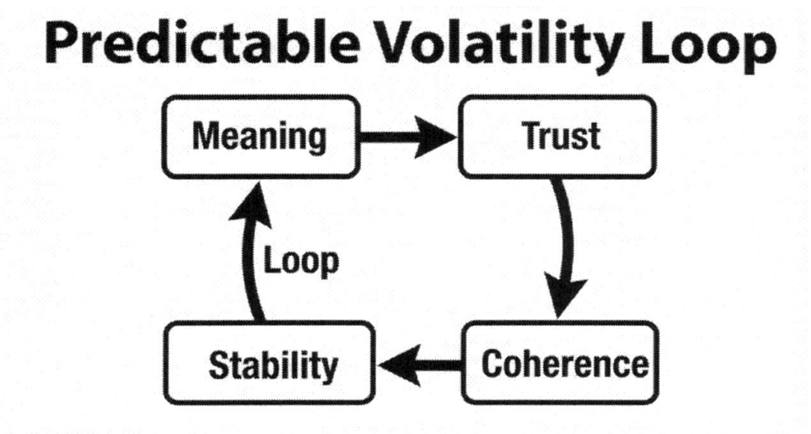

When meaning drifts, the loop does not break.
It continues, enforcing an interpretation that no longer fits reality.

THE ARGUMENT IN THREE SENTENCES

1. Intelligent systems enforce meaning at scale. Once meaning is wrong or outdated, intelligence does not correct it, it accelerates it.
2. The failures documented in this book did not happen because people were careless. They happened because no one was authorised to question what 'right' meant once execution began.
3. Predictable volatility is not fate. It is the consequence of designing systems that value execution over interpretation, coherence over reality, and speed over judgment.

THE FOUR DIAGNOSTIC CONDITIONS

DEFINITION INTEGRITY Can you trace the definitions inside your automated systems back to the conditions they were calibrated for?	CONTEXT PRESERVATION When your systems produce outputs, does the context that gave those outputs their meaning travel with them?
CONTESTABILITY Can a person closest to consequence interrupt execution, without escalation, without retaliation?	DRIFT DETECTION Does a mechanism exist to identify when operating definitions have stopped describing the reality they were built for?

THE QUESTION THAT PRECEDES EVERYTHING ELSE

'When your systems continue to act after meaning has broken, who, exactly, is authorised to stop them?'

THE THREE-STEP RESPONSE

1. IDENTIFY	2. ASK	3. COMMISSION
The system operating at consequential scale where authority to interrupt execution is unclear.	Does this system still mean what we authorised it to mean? If answering requires research, the Diagnostic is overdue.	The Decision Architecture Diagnostic before the next system is authorised or the next deployment is scaled.

> *If that condition exists in your organisation, failure is not a possibility.*
>
> *It is a timing question.*

predictablevolatility.com

MONDAY MORNING

Four actions. No budget. No board paper.
No external engagement required.

FOUR ACTIONS - ONE SYSTEM - ONE MEETING

01 **Name it.** Write down what the system is authorised to decide, not what it does, but what definitions it is operating on and when they were last examined against current conditions. If the date cannot be named, the first gap has already been found.	02 **Ask the Override question.** Ask in your next governance or AI meeting: what is our current override rate and what is it telling us? Not whether it is too high or too low, what it is telling us. An override rate is a signal, not a compliance metric.
03 **Name who has the Right to Stop.** Not who can raise a concern or request a review. Who, specifically, can interrupt execution immediately without personal consequence. If you cannot name that person, the Right to Stop exists in a document, not in your organisation.	04 **Find the Person.** The person closest to consequence, not the boardroom. Ask what they see that is not in the reporting. Then ask: does what they tell you have a pathway to reach someone with authority to act? If it does not, you have found the gap.

These four actions will not fix the problem. They will tell you whether the problem exists and how deep it runs. What follows is the work.

> *Every organisation in this book had people who could have asked these questions.*
>
> *None of them were asked to.*

THE MEANING AUDIT

A diagnostic for organisations deploying AI in consequential decisions. The cases in this book describe organisations that knew something was wrong and continued anyway, not through bad faith, but because no mechanism existed to translate awareness into action. The Meaning Audit is the structured diagnostic that makes that mechanism visible. It is a direct examination of an organisation's Meaning Mesh, the layer between data and decision that most governance frameworks never formally inspect. It surfaces where definitions have expired without revision, where context has been stripped from outputs, where contestability exists only on paper, and where Meaning Drift has accumulated beyond the organisation's awareness.

What a Meaning Audit Examines

The four conditions introduced in Chapter 9 are the audit's diagnostic structure. Definition Integrity examines where the concepts driving consequential AI decisions are explicitly defined, who is authorised to revise them, and when they were last examined against current conditions. Context Preservation examines whether system outputs carry the assumptions and confidence bounds that allow the humans acting on them to understand what they are deciding. Contestability examines whether a formal, protected pathway exists for a member of the organisation to interrupt execution when meaning no longer fits reality, not enter a review cycle, but interrupt. Drift Detection examines whether monitoring tracks whether the meanings the system enforces remain accurate, not only whether performance metrics are within expected ranges.

This framework builds on diagnostic work published in The AI Edge (2026). The original five questions have been restructured here into four interdependent conditions namely, Definition Integrity, Context Preservation, Contestability, and Drift Detection, aligned with the Meaning Mesh architecture. The fifth question, concerning ownership of meaning evolution, is now distributed across conditions one and four.

Who It Is For

The Meaning Audit is designed for boards, CAIOs, CDOs, CROs, CIOs, and executive teams who are preparing to deploy AI systems at consequential scale and want to know whether their decision architecture can support that deployment safely; who are reviewing governance after a near-miss, a regulatory challenge, or a failure that the process failed to anticipate; or who can answer fewer than four of the diagnostic questions in Chapter 9 with confidence and evidence rather than aspiration. It is not a compliance exercise. Organisations that want a compliance report already have the tools for that. The Meaning Audit is for leadership teams that want to understand what their decision architecture looks like before the next system goes live, not after.

The purpose of the Meaning Audit is not compliance certification. It is reducing the probability that organisational confidence diverges silently from operational reality.

How to Engage

Engagements follow a three-step pathway, scaled to what the organisation needs.

A **Decision Diagnostic call** (thirty minutes) establishes whether the four conditions from Chapter 9 can be answered with evidence. Most organisations discover at this stage that one or two conditions exist only in documentation. The call is the starting point, not a sales conversation.

For organisations where the Diagnostic confirms structural exposure, a **Decision Architecture Diagnostic** (half-day session) maps the specific points at which interpretive authority has been removed, contested, or automated without revision. It produces a clear picture of where the decision architecture is sound and where it is not.

Where the Review identifies material gaps, a **Meaning Audit** (four to six weeks, depending on organisational size and system complexity) conducts the full structured examination: definitions reviewed against current conditions, contestability pathways tested for operational reality, drift mechanisms assessed for semantic as well as statistical drift. It establishes, with evidence, whether the systems currently running are operating within their governance boundary.

Conversations begin at predictablevolatility.com

> *The question is not whether to take it seriously. The question is whether to take it seriously before the cost of not doing so arrives.*

CONCEPT LEXICON

A reference guide to the proprietary concepts developed in this book, organised in the sequence in which the argument builds. Terms appear in the order they are encountered across the four stages of the analysis. Chapter references indicate where each term is introduced and defined in full.

STAGE 1: HOW MEANING BREAKS

The conditions under which organisations lose contact with reality while appearing stable

The Predictable Volatility Loop Chapter 1

The four-stage cycle through which meaning, trust, coherence, and stability reinforce one another, and through which their breakdown propagates. The loop describes both the healthy state and the failure mode. The PV Loop is the book's primary diagnostic model for understanding why organisational failures are predictable rather than random.

> *Introduced in Chapter 1 through the Post Office Horizon case and runs as the structural model throughout the book. When the loop breaks at the meaning stage, every subsequent stage compounds the damage.*

Meaning Monopoly Chapter 1, p.6

The condition in which a single system, output, or narrative is granted exclusive authority to define reality within an organisation. Once established, contradictions are resolved automatically in the system's favour. Evidence that challenges the system is treated as suspect because it challenges the system.

> *Horizon's transformation from accounting tool to sole arbiter of truth is the book's primary example. A Meaning Monopoly does not require malice. It requires only a configuration of authority that makes challenge structurally unavailable.*

Interpretation Gap Chapter 2, p.16

The operational shortfall between what a system is doing and what the organisation can understand, contest, and act upon in time. The Interpretation Gap is not a data gap. It is a sensemaking and authority gap that opens whenever execution speed exceeds the organisation's capacity to interpret what is happening and redirect action before consequences compound.

> *Knight Capital's 45-minute loss and the Flash Crash illustrate the Interpretation Gap at different scales: one within a single firm, one across an entire market. In both cases, the system was correct by its own criteria. The gap was between its criteria and the reality it was acting upon.*

Coherent Catastrophe Chapter 2, p.20

The condition in which a failure is simultaneously coherent at every individual level and catastrophic at the systemic level. Each participant performs correctly according to their own logic. The catastrophe emerges from the aggregate of individually correct decisions in the absence of a shared interpretive framework.

The Flash Crash is the clearest example: every HFT algorithm behaved rationally. The market collapsed because no shared semantic protocol governed their collective behaviour. Coherent Catastrophe requires no single point of error.

The Missing Middle Chapter 2, p.162

The interpretive space between system output and human understanding that organisations routinely fail to design. The layer at which raw outputs are translated into actionable meaning. When absent, humans either automate blind or interpret without the information required to interpret accurately.

Knight Capital's operators could see that something was wrong. They could not identify what. The Missing Middle is the gap between 'something is wrong' and 'this specific thing is wrong, and this is what to do about it.'

False Coherence Chapter 3, p.33

The condition in which an organisation's internal narrative remains consistent even as it loses contact with external reality. False Coherence is produced, not accidental. When reality begins to diverge from expectations, the system does not instinctively revise its story. It revises its interpretation of reality to preserve the story.

Wirecard's five-year survival despite abundant contradictory evidence is the book's primary example. False Coherence does not feel like self-deception from the inside. It feels like institutional maturity and responsible management of uncertainty.

Narrative Sinkhole Chapter 3, p.27

The condition in which an organisation's dominant narrative has acquired sufficient gravitational force that questioning it becomes structurally irrational. The organisation no longer asks whether the story is true. It asks whether the story is stable. A Narrative Sinkhole is the advanced form of False Coherence. The costs of maintaining the narrative have been so thoroughly distributed that no individual experiences the full weight of the decision to continue. Everyone acts rationally. The collective produces catastrophe.

Meaning Debt Chapter 3, p.31

The growing gap between an organisation's narrative and the reality it is operating in. Meaning Debt is not visible on financial statements. But it compounds. The longer the gap is preserved, the more destructive the eventual correction becomes.

> *Volkswagen's Clean Diesel programme accumulated Meaning Debt across each model year. The collapse appeared sudden. In reality, it was the delayed settlement of Meaning Debt that had been accumulating for years.*

Meaning Mesh Breakdown Chapter 3, p.28

The diagnostic framework that maps where meaning is being generated, contested, assumed, and suppressed within a system under epistemic stress. When the diagonal between the Evidence Test and the Institutional Narrative is broken, the system can no longer self-correct.

> *Applied forensically in Case 5 (Dialogue of the Deaf) to the Post Office Horizon conflict. The framework is diagnostic rather than prescriptive; it locates where meaning has broken before prescribing how to repair it.*

Human Signal Chapter 4, p.40

The direct experiential testimony of the person closest to a consequence: the operator who sees the anomaly, the engineer who identifies the constraint, the frontline worker who observes the discrepancy. The earliest available indicator of system failure. Also the signal that organisational authority structures are most systematically designed to discount.

> *Roger Boisjoly's July 1985 memo is the clearest primary record of a Human Signal: explicit, documented, prophetic, and ignored. The Human Signal is not always expressed in the language authority understands. This is not its weakness. It is its nature.*

Threshold Moment Chapter 4, p.42

The final point at which judgment is structurally available, authority is still recoverable, and course correction remains proportionate. Once passed, reversal becomes organisationally illegitimate, not because the evidence changes, but because the system does. The Threshold Moment is only identifiable in retrospect, which is precisely what makes it dangerous.

> *The teleconference on January 27, 1986 was a Threshold Moment. It passed quietly. Then Challenger launched.*

Interpretive Demotion Chapter 5, p.51

The organisational mechanism by which constraint-based interpretation is recategorised from strategic signal to attitude problem. The content of the concern is not refuted; its institutional status is changed, placing it beyond the reach of decision-making authority. Two forms appear: goal-protection demotion (Volkswagen, Challenger) and futility demotion (Columbia). In both cases, the valid signal was reclassified without confrontation, requiring only a change in the interpretive frame through which it was received.

STAGE 2: HOW ORGANISATIONS SUSTAIN THE BREAK

The mechanisms through which awareness is decoupled from the authority to act

Managed Knowing Chapter 6, p.63

The condition in which an organisation possesses accurate awareness of a developing failure but has structurally decoupled that awareness from the authority to act. The organisation is neither uninformed nor in denial, it knows, and continues.

> *Boeing's internal awareness of MCAS's single-sensor dependency is the book's primary evidence. The 2016 engineer's communication, 'running like a runaway train', was not suppressed. It was processed into inaction. This is not ignorance. It is Managed Knowing.*

Meaning Drift Chapter 8, p.96

The slow divergence of an interpretation from the reality it was designed to describe, occurring without announcement or formal decision. Definitions that were accurate when established gradually cease to describe the conditions they were calibrated for.

> *The language of structured finance in the years before 2008 is the book's primary example: not falsified but drifted. Each abstraction step moved the vocabulary further from the underlying reality. Meaning Drift is invisible because the language remains stable. It is the reality beneath the language that moves.*

Narrative Capture Chapter 7, p.71

The condition in which a dominant organisational narrative has become so deeply embedded in institutional structures, incentive systems, and professional identities that it can no longer be evaluated from within. Narrative Capture is not a conspiracy. It is an emergent property of sufficiently aligned incentive structures.

> *The Global Financial Crisis demonstrates Narrative Capture at systemic scale: analysts, auditors, regulators, and investors all operating within a narrative that rewarded alignment and penalised challenge.*

Contestability Chapter 7, p.74

The operational condition in which an organisation's meanings, classifications, and decisions remain open to challenge from within. Distinct from governance structure: an organisation can have every formal governance feature while remaining wholly uncontestable. Without Contestability, governance becomes performance.

> *Toshiba had exemplary formal governance. It had no Contestability. The cultural architecture ensured that challenge was structurally impossible regardless of the formal mechanisms available.*

Absurdity Audit Chapter 7, p.75

A practical diagnostic in which a sentence from a board deck, policy memo, or strategy document is translated into ordinary language. If the translation preserves meaning but changes tone, the abstraction was functional. If the translation reveals a different reality, if the plain version contains a fact the corporate version obscures, the organisation is inside a Narrative Sinkhole.

> *'The model's outputs are subject to continuous human oversight and quality assurance processes' translated as 'a reviewer checks a sample of decisions after they have been executed and flags patterns to a team whose recommendations are advisory.' The discomfort produced by an accurate translation is not a flaw in the diagnostic. It is the diagnostic working.*

STAGE 3: WHAT AUTOMATION DOES TO THE BREAK

The specific properties that intelligent systems add to existing meaning failures

Interpretive Collapse Chapter 5, p.52

The condition in which multiple actors derive different meanings from the same system output and proceed as if aligned, each operating on their own interpretation without awareness of the divergence.

> *Boeing's MCAS is the book's primary example: engineers, pilots, and regulators all held different definitions of what MCAS was. None corresponded to what it actually did. Everyone believed they understood the same system. They were not.*

Frozen Meaning Chapter 8, p.96

The condition in which an interpretation has been embedded so deeply into automated systems that it can no longer be revised without halting execution. Where Vaughan's normalised deviance could still be renegotiated by humans, Frozen Meaning has removed the renegotiation mechanism entirely. The deviance no longer normalises. It institutionalises.

Robodebt's income averaging method, Toeslagenaffaire's encoding of dual nationality as a fraud-risk proxy, Zillow's Zestimate as acquisition authority, and MCAS's authority expansion are all examples of meaning frozen at a point in time and then executed without the possibility of human renegotiation. Deep Thought's answer of 42, correct, faithfully produced, and operationally meaningless without the Question, is the literary archetype.

STAGE 4: THE DIAGNOSTIC AND CORRECTIVE ARCHITECTURE

The structural conditions required for organisations to remain interpretively resilient

Meaning Mesh Chapter 9, p.105

The distributed layer of definitions, assumptions, narratives, incentives, and permissions that sits between data and action. The Meaning Mesh is not a document, a model, or a policy. It determines what counts as a signal, how ambiguity is resolved, which definitions are treated as fixed, and who is authorised to pause execution. When invisible and unexamined, the Meaning Mesh is the primary site of organisational meaning failure. Making it visible is the first act of interpretive resilience.

Decision Architecture Chapter 8, p.92

The practical arrangement of roles, rights, evidence, escalation paths, accountability, and stopping power that determines how meaning becomes action inside an organisation. Decision Architecture determines who can interpret, decide, challenge, pause, stop, or redirect action, especially when automated systems are operating faster than human sensemaking can comfortably follow.

The Horizon case, Knight Capital, Robodebt, and Boeing all illustrate Decision Architecture failures in different forms. In each case, the instrument of correction existed, but the authority structure surrounding it ensured it could not function when needed.

Dual-Key Interpretation Chapter 9, p.109

A structural design principle requiring that no irreversible decision proceeds unless two independent conditions are simultaneously met: the Authority Key (formal system authorisation) and the Proximity Key (human confirmation from the party closest to consequence). If these keys cannot both be engaged, execution pauses. The pause is not inefficiency. It is the organisation thinking.

Dual-Key Interpretation is the minimum corrective design for the condition in which humans are present but authority is absent, the condition that produced Challenger, the Horizon prosecutions, and the Robodebt debt notices.

Interpretive Resilience Chapter 9, p.97

The structural condition in which an organisation's meaning systems remain open to revision even under the pressure of scale, speed, and automation. A design property, the outcome of deliberately maintaining Definition Integrity, Context Preservation, Contestability, and Drift Detection across the organisation's decision architecture. Interpretive Resilience is the book's answer to the question implicit in every case: what would have had to be true for this failure not to have occurred? In every case, its absence was structural rather than personal.

The Right to Stop Chapter 9, p.103

The formal, protected authority of a designated individual or group to halt automated execution when meaning no longer fits the reality being acted upon. Not the right to raise a concern. The right to interrupt, immediately, without escalation, without personal consequence.

> *HAL 9000's 'I'm sorry Dave, I can't do that' is the cultural image of a system from which the Right to Stop has been removed, inverted: the machine will not let the human stop it. The CME's five-second Flash Crash pause demonstrates what the Right to Stop looks like at minimal design.*

Meaning Audit Chapter 9, p.106

A structured diagnostic of an organisation's Meaning Mesh, designed to surface where definitions have expired without revision, where context has been separated from outputs, where contestability is decorative rather than operational, and where Meaning Drift has accumulated undetected. The Meaning Audit is the book's primary commercial instrument. The four diagnostic questions in Chapter 9 are the entry point. The note at the back of the book is where to start.

The Evocati Epilogue, p.132

In Roman armies, the Evocati were veterans recalled to serve. They were experienced, respected, and often closest to what was actually happening. But they no longer held command. They could see. They could warn. They could not stop what had already been set in motion. Defined not by role or seniority but by the willingness to name what they see when naming it is professionally inconvenient. The Epilogue positions the Evocati as the human answer to the structural argument the book makes. If systems cannot self-correct, and if governance structures can be captured, the corrective function must rest somewhere in human judgment. The Evocati are the people capable of exercising it.

Using This Lexicon

Terms in this lexicon are tools for diagnosis, not classification. Their value is not in applying labels to situations already understood, but in surfacing conditions that organisations typically do not have the vocabulary to name and therefore cannot address. A Meaning Monopoly does not feel like a Meaning Monopoly from the inside. Managed Knowing does not feel like a structural failure. False Coherence does not feel false.

That is precisely why these conditions persist. Naming them does not resolve them. But it makes resolution possible. These are the conditions the book documents. The Forensic Cases show what they cost. The Meaning Audit is where the diagnosis begins.

LIST OF FIGURES

REFERENCES

Primary sources for the forensic cases, academic foundations, and regulatory records cited in this book.

Academic and Intellectual Foundations

Argyris, C. Organisational Learning: A Theory of Action Perspective. Addison-Wesley, 1978.

Argyris, C. and Schön, D.A. Organisational Learning II: Theory, Method, and Practice. Addison-Wesley, 1996.

Edmondson, A.C. The Fearless Organisation: Creating Psychological Safety in the Workplace for Learning, Innovation, and Growth. Wiley, 2018.

Edmondson, A.C. 'Psychological Safety and Learning Behaviour in Work Teams.' Administrative Science Quarterly 44(2), 1999, pp. 350–383.

Obermeyer, Z., Powers, B., Vogeli, C., and Mullainathan, S. "Dissecting racial bias in an algorithm used to manage the health of populations." Science, 366(6464), 2019, pp.447–453. DOI: 10.1126/science.aax2342.

Simonite, T. "A Health Care Algorithm Offered Less Care to Black Patients." WIRED, 24 October 2019.

Vaughan, D. The Challenger Launch Decision: Risky Technology, Culture, and Deviance at NASA. University of Chicago Press, 1996.

Vaughan, D. 'The Normalization of Deviance: Signals and Rules in Organisational Risk.' Annual Review of Sociology 25, 1999, pp. 271–305.

Weick, K.E. Sensemaking in Organisations. Sage, 1995.

Weick, K.E. The Social Psychology of Organizing. Addison-Wesley, 1979.

Weick, K.E., Sutcliffe, K.M. and Obstfeld, D. 'Organizing and the Process of Sensemaking.' Organisation Science 16(4), 2005, pp. 409–421.

Bengio, Y. et al. International AI Safety Report 2026. OECD/UN/EU, 2026.

Post Office Horizon

Bates v Post Office Ltd (No.6: Horizon Issues) High Court of Justice, Business and Property Courts, 2019. EWHC 3408 (QB).

Horizon IT Inquiry Transcripts, interim reports, and witness evidence. UK Statutory Inquiry, 2021–present. www.horizonitinquiry.org.uk.

Court of Appeal Hamilton and others v Post Office Ltd. [2021] EWCA Crim 577. Quashing of convictions on grounds of abuse of process.

Post Office Act 2024. UK Parliament. Legislation providing statutory exoneration for wrongfully convicted sub postmasters.

Boeing 737 MAX

U.S. House Committee on Transportation and Infrastructure. The Design, Development, and Certification of the Boeing 737 MAX. Final Committee Report, September 2020.

Joint Authorities Technical Review (JATR). Observations, Findings, and Recommendations to the Federal Aviation Administration. October 2019.

National Transportation Safety Board (NTSB). Assumptions Used in the Safety Assessment Process and the Effects of Multiple Alerts and Indications on Pilot Performance. Safety Recommendation Report, September 2019.

Global Financial Crisis 2007–2009

Financial Crisis Inquiry Commission (FCIC). The Financial Crisis Inquiry Report. U.S. Government Publishing Office, January 2011.

Turner, A. The Turner Review: A Regulatory Response to the Global Banking Crisis. Financial Services Authority, March 2009.

U.S. Senate Permanent Subcommittee on Investigations. Wall Street and the Financial Crisis: Anatomy of a Financial Collapse (Levin-Coburn Report). April 2011.

Parliamentary Commission on Banking Standards. Changing Banking for Good. House of Lords/House of Commons, June 2013.

Volkswagen Diesel Emissions

U.S. Department of Justice. United States of America v. Volkswagen AG. Plea agreement and statement of facts, January 2017. Case No. 16-cr-20394.

U.S. Environmental Protection Agency. Notice of Violation - Clean Air Act. September 18, 2015.

Braunschweig Regional Court (Landgericht Braunschweig). Proceedings against Volkswagen senior executives. Ongoing from 2020.

Toshiba Corporation

Toshiba Corporation Independent Investigation Committee. Report of the Investigation Committee on the Accounting Issues at Toshiba Corporation. July 20, 2015.

Financial Services Agency (Japan). Administrative sanctions and enforcement record. 2015.

Toshiba Independent Investigation Panel on Shareholders' Meeting. Investigation Report. November 2021.

Wirecard AG

German Federal Parliament (Bundestag). Report of the Parliamentary Investigation Committee on the Wirecard Scandal. 2020–2021.

Munich Public Prosecutor's Office. Criminal indictments against Markus Braun and co-defendants. 2022-2023.

Zillow Offers

Zillow Group, Inc. Quarterly Report on Form 10-Q for the period ended September 30, 2021. U.S. Securities and Exchange Commission filing, November 2021.

Zillow Group, Inc. Q3 2021 Earnings Call Transcript. November 2, 2021.

Robodebt (Online Compliance Intervention)

Royal Commission into the Robodebt Scheme. Final Report. Commonwealth of Australia, July 2023.

Gordon Legal and Alden Compensation. Robodebt Class Action Settlement. Federal Court of Australia, November 2021.

Amato v Commonwealth of Australia. [2020] FCA 1 - Federal Court of Australia ruling on unlawfulness of income averaging debts.

Toeslagenaffaire

Parliamentary Inquiry Committee Report Ongekend Onrecht (Unprecedented Injustice), December 2020. Tweede Kamer der Staten-Generaal, The Hague.

Adviescommissie Uitvoering Toeslagen (Van Dam Commission) Reports, 2020. Independent review of the Tax Administration's handling of benefit recovery.

Autoriteit Persoonsgegevens, Investigation Report into the use of nationality as a fraud-risk indicator in automated screening, 2021.

Knight Capital Group / Flash Crash

U.S. Securities and Exchange Commission. Administrative Proceeding, File No. 3-15570 - In the Matter of Knight Capital Americas LLC. October 16, 2013.

U.S. Securities and Exchange Commission and Commodity Futures Trading Commission. Findings Regarding the Market Events of May 6, 2010. Report of the Staffs of the CFTC and SEC, September 30, 2010.

Space Shuttle Challenger (1986)

Presidential Commission on the Space Shuttle Challenger Accident (Rogers Commission). Report to the President. June 6, 1986.

Feynman, R.P. 'Personal Observations on the Reliability of the Shuttle.' Appendix F to the Rogers Commission Report, 1986.

Space Shuttle Columbia (2003)

Columbia Accident Investigation Board (CAIB). Report, Volume 1. National Aeronautics and Space Administration, August 2003.

FURTHER READING

Selected works for readers who want to go deeper on the arguments, cases, and intellectual lineage of this book.

On Organisational Failure and Sensemaking

Weick, K.E. Sensemaking in Organisations. Sage, 1995. The foundational work on how people construct meaning under uncertainty. Essential for understanding the Interpretation Gap.

Edmondson, A.C. The Fearless Organisation. Wiley, 2018. The definitive account of psychological safety and why people do not speak up. Directly relevant to every chapter on signal suppression.

Vaughan, D. The Challenger Launch Decision. University of Chicago Press, 1996. The most rigorous forensic account of normalisation of deviance available. Vaughan's framework is the intellectual predecessor to Frozen Meaning.

Vaughan, D. Dead Reckoning: Air Traffic Control, System Effects, and Risk. University of Chicago Press, 2021. Vaughan's extension of normalisation of deviance into complex sociotechnical systems. The intellectual context for the Frozen Meaning argument in Chapter 8.

Perrow, C. Normal Accidents: Living with High-Risk Technologies. Basic Books, 1984. The foundational work on systemic failure in tightly coupled systems.

Reason, J. Managing the Risks of Organisational Accidents. Ashgate, 1997. The Swiss cheese model and the latent failure conditions that precede catastrophe.

On Decision-Making and Judgment Under Uncertainty

Kahneman, D. Thinking, Fast and Slow. Farrar, Straus and Giroux, 2011. The cognitive architecture beneath every case of judgment being systematically overridden.

Duke, A. Thinking in Bets: Making Smarter Decisions When You Don't Have All the Facts. Portfolio/Penguin, 2018. The pre-mortem as a contestability mechanism, the most accessible treatment of probabilistic decision-making for a practitioner audience.

Klein, G. Sources of Power: How People Make Decisions. MIT Press, 1998. Recognition-primed decision-making and the expertise that produces the Human Signal.

On AI Risk and Algorithmic Accountability

O'Neil, C. Weapons of Math Destruction: How Big Data Increases Inequality and Threatens Democracy. Crown, 2016. The most accessible forensic account of how algorithms encode and scale human biases. Directly relevant to Chapters 8 and 10.

Pasquale, F. The Black Box Society: The Secret Algorithms That Control Money and Information. Harvard University Press, 2015. The legal and governance dimensions of algorithmic opacity.

Eubanks, V. Automating Inequality: How High-Tech Tools Profile, Police, and Punish the Poor. St. Martin's Press, 2018. The structural harm of automated welfare decisions, the intellectual companion to the Robodebt case.

On Language, Power, and Institutional Failure

Orwell, G. Nineteen Eighty-Four. Secker & Warburg, 1949. Newspeak, doublethink, and the memory hole - Orwell's diagnostic of how language is engineered to make dissent unthinkable. The literary ancestor of every chapter on linguistic anaesthesia.

Orwell, G. 'Politics and the English Language.' Horizon, April 1946. The essay that named the mechanism: language used not to illuminate but to insulate authority from challenge.

Adams, D. The Hitchhiker's Guide to the Galaxy. Pan Books, 1979. Deep Thought, the Answer of 42, and the Sirius Cybernetics Corporation: Adams understood the absurdity of systems executing faithfully in contexts that had ceased to exist. Required reading for Chapter 8.

Adams, D. The Restaurant at the End of the Universe. Pan Books, 1980. The Sirius Cybernetics Corporation's complaints department buried underground. The definitive comic treatment of Contestability as governance theatre.

On Satire as Diagnostic

Chapman, G., Cleese, J., Gilliam, T., Idle, E., Jones, T. and Palin, M. Monty Python and the Holy Grail. Python (Monty) Pictures, 1975. The Black Knight as False Coherence made visible: acknowledgment without revision, executed with perfect comic timing. Relevant to Chapter 3.

Kubrick, S. (director) and Clarke, A.C. (screenplay). 2001: A Space Odyssey. MGM/Stanley Kubrick Productions, 1968. HAL 9000's 'I'm sorry Dave, I can't do that' is the definitive cultural image of a system that has removed the Right to Stop. Relevant to Chapter 8 and Chapter 9.

On Financial Markets and Systemic Risk

Lewis, M. The Big Short: Inside the Doomsday Machine. W.W. Norton, 2010. The most readable account of how linguistic engineering and incentive alignment produced the GFC.

Tett, G. Fool's Gold: How the Bold Dream of a Small Tribe at J.P. Morgan Was Corrupted by Wall Street Greed and Unleashed a Catastrophe. Free Press, 2009. The anthropological account of how financial innovation produced the conditions for False Coherence at systemic scale.

On the Practice of Decision Architecture

Turner, A.J. Predictable Volatility: Why Meaning, Not Models, Determines Whether AI Breaks You or Builds You. Foundarity OÜ, 2026. The primary source for the Meaning Audit, Dual-Key Interpretation, and the Decision Architecture framework developed in this book.

Turner, A.J. 'Predictable Volatility Series.' Substack, 2026 onwards. https://predictablevolatility.substack.com.

Turner, A.J. 'Predictable Volatility: Why Meaning, Not Models, Determines Whether AI Breaks You or Builds You.' In Seversen, E. (ed.), The AI Edge. Thin Leaf Press, 2026. The chapter that introduced the core framework to a practitioner audience.

ACKNOWLEDGEMENTS

Books are written alone. They are not made alone.

This one emerged from more than thirty years of observation inside organisations trying, in most cases, to do the right thing under conditions that made meaning difficult to preserve. The people acknowledged here did not simply support the writing of this book; they pressure-tested its claims, challenged its blind spots, and sharpened the argument where it was weakest.

Predictable Volatility is a forensic account of structures, not individuals. But it exists because of individuals.

The intellectual foundations beneath this work rest heavily on Diane Vaughan's concept of the normalisation of deviance, alongside the contributions of Karl Weick, Chris Argyris, and Amy Edmondson. Their work runs beneath almost every chapter, even where they are not explicitly named.

Debbie Bayntun-Lees read the manuscript with exceptional rigour and generosity, identifying where the argument needed to earn what it was asserting and where the diagnostic framework had to become more teachable and operationally usable. Roberto Maranca approached the work as a practising data executive rather than an abstract theorist, confirming that the framework holds where it matters most: inside operational decision-making realities.

Jon Hammant pressed hardest on the book's architectural claims, insisting that the distinction between design, operation, and accountability be made explicit rather than implied. Several arguments in Chapter 9 are sharper because of that pressure.

Ian Gotts identified where the manuscript was losing readers before the argument had fully landed, leading to structural changes in pacing, diagram placement, and chapter sequencing that materially improved accessibility. Dave Snowden pushed the framework toward its complexity edge cases, particularly around distributed sense-making and the conditions under which interpretation collapses collectively rather than individually. Tony Fish challenged the argument from a governance and fiduciary accountability perspective, pressing on the legal and board-level consequences of the failure modes described throughout the book.

Many of the practitioners who engaged with this manuscript did so while carrying operational accountability inside complex systems under real-world pressure. Their engagement with the argument was not academic. In many cases, they recognised patterns they had already lived through.

Rowland Agidee brought the perspective of deploying AI in acute healthcare environments, while Campbell McLean confirmed independently, from an infrastructure and operations perspective, that the book's central claim holds most visibly where meaning failure has immediate operational consequence. Peter Depledge recognised the framework's precision from direct experience across nuclear, offshore energy and major capital programmes, environments where the absence of the 'so what' question has consequences that are neither abstract nor recoverable. Peter Ward applied the argument from a CFO and board accountability perspective, identifying Chapter 10 in particular as essential reading for any leader assuming that automated design can be trusted without sustained interrogation. Valerie Logan and Russell Miles both recognised the unresolved frontier beneath the manuscript: how systems might be designed so meaning remains contestable rather than frozen.

Olga Martynov validated that the argument holds not only technically, but at board and audit committee level, while Shelley Pittman recognised in the forensic cases patterns financial services practitioners have watched emerge for years without a common language to describe them.

Joe Toscano reinforced the book's central distinction: that the problem is rarely the technical system alone, but the interpretation layer surrounding it. Bruno Goffin confirmed that courageous leadership and structural conditions are complements, not alternatives. Without contestability designed into the system, courage is often routed rather than heard. Hendo brought the creative industry perspective the framework had not explicitly addressed, recognising that the Meaning Mesh has operated informally inside creative and brand environments for decades. Christopher Lochhead identified the problem frame itself as one of the manuscript's most important contributions.

The collaboration that began around The AI Edge accelerated the momentum behind finishing this book. Erik Seversen consistently reinforced the importance of allowing the work to develop on its own terms rather than reshaping it to fit market convention. That encouragement made a material difference in the final stages of the manuscript.

Simeon Rad, Akhil Lalwani, and Alasdair Anderson gave the manuscript a careful late-stage read-through at a critical moment. Their observations on clarity and register improved the final text considerably.

Morgane Calenca challenged my thinking on how organisational failure is measured and made visible. Her contribution to the Cost of Failure analysis sharpened both the framework and the table that presents it.

Several conversations on The G&T Sessions podcast provided the earliest sparks for ideas that eventually became chapters in this book. The practitioners, leaders, and thinkers who engaged with those discussions formed the first audience for arguments that are now in print.

Finally, to everyone who raised a concern that was routed rather than heard, who asked the question that ended the meeting, who was right too early and paid for it: this book was written because of what happened to you.

The structures it describes are not inevitable. They are designed. And what is designed can be redesigned.

Andrew J. Turner
London, 2026

ABOUT THE AUTHOR

Andrew J. Turner

Author • Advisor • Founder

Andrew J. Turner has operated inside some of the world's most complex organisations, GE, SAP, Telefónica, and Tesco, latterly as COO of Tesco Mobile. In the years since, he has worked as a founder, operator, and senior advisor across financial services, technology, and regulated industries, most recently focused on AI governance and decision risk, work that produced the framework in this book. He works with leadership teams and boards on Decision Diagnostics, Decision Architecture and Meaning Audits; the practical instruments developed in this book. He advises founders, investors, and executive teams on AI readiness and decision risk in complex, high-stakes environments.

He read Nineteen Eighty-Four at sixteen and spent thirty years watching organisations build the architecture Orwell described without the totalitarian apparatus, through incentive alignment, linguistic anaesthesia, and the structural suppression of dissent. That observation is the intellectual spine of this book. Andrew is a Contributing Author to The AI Edge (Thin Leaf Press, 2026), in which the Predictable Volatility framework made its first public appearance. He holds an MBA from Ashridge Business School and hosts The G&T Sessions podcast.

Web: predictablevolatility.com

LinkedIn: linkedin.com/in/iamandrewjturner

DID YOU ENJOY THIS BOOK?

If you enjoyed reading this book, you can help by suggesting it to someone else you think might like it, and **please leave a positive review** wherever you purchased it. This does a lot in helping others find the book. We thank you in advance for taking a few moments to do this.

THANK YOU

INDEX

A

B

C

D

E

F

G

H

I

K

L

M

N

O

P

R

S

T

V

W

Z

Made in United States
Orlando, FL
04 November 2026

90201354R00208